THINKING WOMEN

A TIMELINE OF SUFFRAGE IN UTAH

KATHERINE KITTERMAN & REBEKAH RYAN CLARK

20 | BETTER DAYS | 20

Visit us at deseretbook.com

Library of Congress Cataloging-in-Publication Data

(CIP data on file)
ISBN 978-1-62972-695-3

Printed in China
RR Donnelley, Dongguan, China

10 9 8 7 6 5 4 3 2 1

VOTES FOR WOMEN

"I BELIEVE IN WOMEN, ESPECIALLY THINKING WOMEN."

EMMELINE B. WELLS

CONSTITUTION

OF THE

NATIONAL WOMAN SUFFRAGE ASSOCIATION.

ARTICLE 1.—This organization shall be called the NATIONAL WOMAN SUFFRAGE ASSOCIATION.

ARTICLE 2.—The object of this Association shall be to secure NATIONAL protection for women citizens in the exercise of their right to vote.

ARTICLE 3.—All citizens of the United States subscribing to this Constitution, and contributing not less than one dollar annually, shall be considered members of the Association, with the right to participate in its deliberations.

ARTICLE 4.—The Officers of this Association shall be a President, a Vice-President from each of the States and Territories, Corresponding and Recording Secretaries, a Treasurer and an Executive Committee of not less than five.

ARTICLE 5.—A quorum of the Executive Committee shall consist of nine, and all the Officers of this Association shall be *ex-officio* members of such Committee, with power to vote.

ARTICLE 6.—All Women Suffrage Societies throughout the country shall be welcomed as auxiliaries; and their accredited officers or duly appointed representatives shall be recognized as members of the National Association.

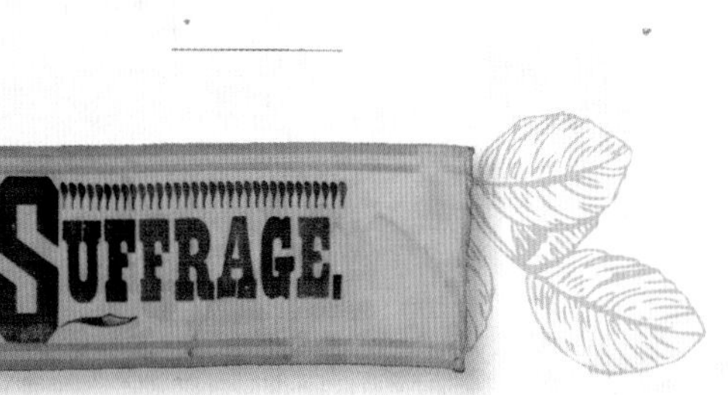

No. 25

CERTIFICATE OF MEMBERSHIP

OF THE

Woman Suffrage Association

OF

SEVIER COUNTY, UTAH,

This is to Certify that [illegible] having subscribed and paid for membership in the Woman Suffrage Association of Sevier County, is entitled to all its benefits and advantages for one year from the date hereof.

[illegible] Secretary. [illegible] President.

DATE [illegible]

N. B.—This Certificate is not Transferable.

Local Membership

WORLD'S

Congress of Representative Women

UNDER THE AUSPICES OF THE

WOMAN'S BRANCH OF THE WORLD'S CONGRESS AUXILIARY

WORLD'S COLUMBIAN EXPOSITION.

MRS. POTTER PALMER, PRESIDENT, MRS. CHARLES HENROTIN, VICE-PRESIDENT.

MEMORIAL ART PALACE,

May 15 to 21, inclusive,

1893.

DEPARTMENT CONGRESS

OF THE

Young Ladies' National Mutual Improvement

ASSOCIATION

Improvement our Motto, Perfection our Aim.

HEADQUARTERS, SALT LAKE CITY, UTAH.

Department Hall No. VII, of the Memorial Art Palace,

MICHIGAN AVENUE, FACING ADAMS STREET, NEAR THE CENTRE OF CHICAGO

FRIDAY, MAY 19, 1893.

EVENING SESSION, 7:45 O'CLOCK.

OFFICERS:

ELMINA S. TAYLOR, PRESIDENT,
MARIA Y. DOUGALL, FIRST VICE-PRES[illegible]
MARTHA H. TINGEY, [illegible]
ANN M. CANNON, [illegible]
MA[illegible]

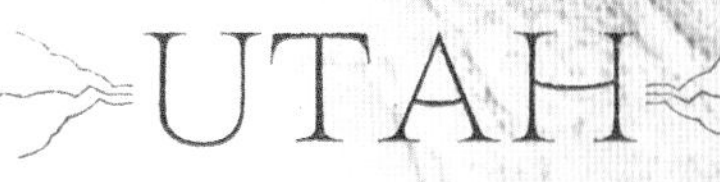

UTAH WOMAN SUFFRAGE

PRESIDENT WOODRUFF'S

MANIFESTO.

PROCEEDINGS AT THE SEMI-ANNUAL GENERAL CONFERENCE OF THE CHURCH OF JESUS CHRIST OF LATTER-DAY SAINTS,

Monday Forenoon, October 6, 1890.

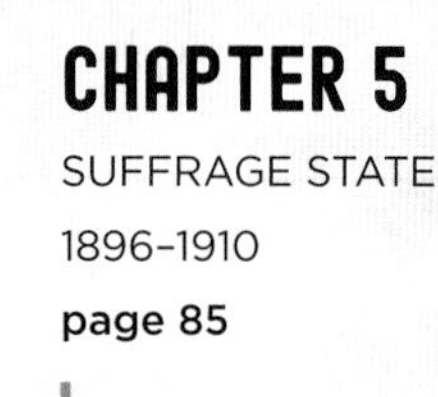

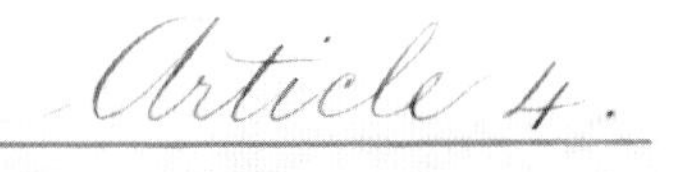

Article 4.

Elections and Right of Suffrage.

Section 1. The rights of citizens of the State of Utah to vote and hold office shall not be denied or abridged on account of sex. Both male and female citizens of this State shall enjoy equally all civil, political and religious rights and privileges.

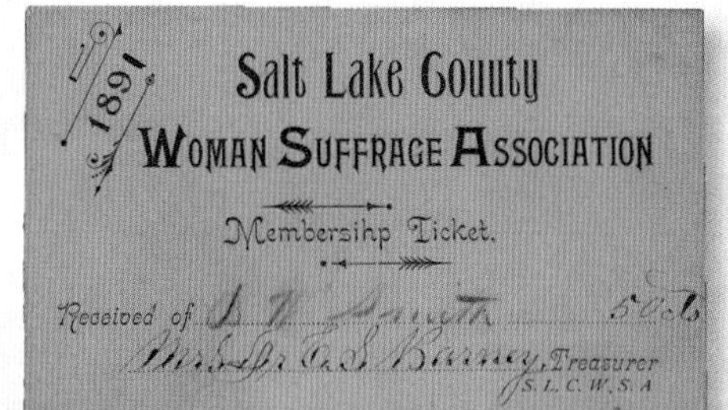

Sixty-sixth Congress of the United States of America;

At the First Session,

Begun and held at the City of Washington on Monday, the nineteenth day of May, one thousand nine hundred and nineteen.

JOINT RESOLUTION

Proposing an amendment to the Constitution extending the right of suffrage to women.

Resolved by the Senate and House of Representatives of the United States of America in Congress assembled (two-thirds of each House concurring therein), That the following article is proposed as an amendment to the Constitution, which shall be valid to all intents and purposes as part of the Constitution when ratified by the legislatures of three-fourths of the several States.

"ARTICLE ——.

"The right of citizens of the United States to vote shall not be denied or abridged by the United States or by any State on account of sex.

"Congress shall have power to enforce this article by appropriate legislation."

Speaker of the House of Representatives.

Vice President of the United States and President of the Senate.

FOREWORD

—BY CHRISTINE M. DURHAM, FORMER CHIEF JUSTICE OF THE UTAH SUPREME COURT

THIS BOOK TELLS, IN NUANCED and well-documented detail, the remarkable story of Utah women and men who, ahead of their time, refused to acquiesce in outdated notions of justice and equality for women; and who undertook a nationally observed and complicated battle for the right to vote.

The story includes the often-overlooked facts that in 1870 the Utah Territory became the second in the nation (after Wyoming a few months earlier) to pass women's suffrage and that in the same year, Utahn Seraph Young became the first woman in the nation to cast a ballot with unrestricted suffrage rights. The story celebrates and preserves a heritage that Utah women may understand both as a source of pride and as an inspiration to devote themselves to contemporary struggles for liberty, justice, and equality.

There is, however, another dimension to the story, having to do with the constitutional structure of the United States and the degree to which the states, through their law-making and particularly through their constitution-making, have contributed to national progress. By the time the Nineteenth Amendment to the federal constitution went into effect in 1920, fifteen states had extended suffrage to women in their state constitutions. Progress at the state level was the result of a deliberate strategy to embed the right to vote in the American experience and to reduce resistance to the idea of women voting. The strategy led to increasing numbers of women with political experience able and willing to join the national fight—many Utah women included.

In 1987, at the beginning of a three-year celebration of the bicentennial of the United States Constitution, U.S. Supreme Court Justice and renowned civil rights advocate Thurgood Marshall offered some startling commentary on the event:

> *"I do not believe that the meaning of the Constitution was forever 'fixed' at the Philadelphia Convention. . . . To the contrary, the government they devised was defective from the start, requiring several amendments, a civil war, and momentous social transformation to attain the system of constitutional government, and its respect for the individual freedoms and human rights, we hold as fundamental today. When contemporary Americans cite 'The Constitution,' they invoke a concept that is vastly different from what the Framers barely began to construct two centuries ago."*

Justice Marshall focused on the meaning of the first three words of the Constitution's Preamble: "We the People." In 1787, of course, this language did not include the majority of the nation's inhabitants—excluding most notably slaves and arguably women, who had no voice in the new polity.

> *"Along the way, new constitutional principles have emerged to meet the challenges of a changing society. . . . The men who gathered in Philadelphia in 1787 could not have envisioned these changes. They could not have imagined, nor would they have accepted, that the document they were drafting would one day be construed by a Supreme Court to which had been appointed a woman and the descendent of an African slave. 'We the People' no longer enslave, but the credit does not belong to the Framers. It belongs to those who refused to acquiesce in outdated notions of*

'liberty,' 'justice,' and 'equality,' and who strived to better them."
—Thurgood Marshall, Bicentennial Speech (1987)

The epilogue of this book notes, perhaps with some of the same sensibilities that Thurgood Marshall brought to the U.S constitutional bicentennial celebration, that our own celebration of the Nineteenth Amendment and of Utah women's suffrage should include reflection on things not done and some promises still unkept. In addition to the voting injustices imposed in this country on people of color and immigrants over many years, the women's suffrage movement itself contains a sad history of indifference and even hostility to the claims of black American women to their rightful place at the table. We must take our pride in and our inspiration from the amazing women who came before us and use their examples to do even better. Justice, liberty, and equality do not exist and endure without human dedication.

Salt Lake NAACP Chapter President Alberta Henry at a Utah Democratic Convention.

UTAH WOMEN LED BY EMMELINE B. WELLS WELCOME NATIONAL SUFFRAGE ENVOYS ON THE STEPS OF THE UTAH STATE CAPITOL, OCTOBER 16, 1915.

INTRODUCTION

"THE STORY OF THE STRUGGLE FOR WOMAN'S SUFFRAGE IN UTAH IS THE STORY OF ALL EFFORTS FOR THE ADVANCEMENT AND BETTERMENT OF HUMANITY."

—DR. MARTHA HUGHES CANNON

UTAH WOMEN LED THE WAY FOR women's rights in many respects, but their suffrage story has largely been lost to history. Most people know about national suffrage leaders like Susan B. Anthony and Elizabeth Cady Stanton, but they are less aware of the large-scale, grassroots effort that the struggle for women's suffrage entailed. It was the petitions, protests, and persuasions of millions of women (and men) that secured voting rights for women across the country. The cumulative effort of everyday people turned the tide of history.

This book tells Utah's suffrage story through the words and actions of the women who first exercised equal suffrage a full fifty years before the passage of the Nineteenth Amendment. Utah women's experience was unique within the national suffrage movement—in part because they secured their rights with the cooperation and support of Utah men and the leadership of the predominant church. This model of women and men working together for the good of the community is inspiring, intriguing, and thought-provoking even today.

American women have always played important roles in their communities, but it took almost a century and a half from the time of the Revolutionary War for women to gain voting rights on a national scale. When the U.S. Constitution was ratified in 1787, it gave states the power to regulate their own voting laws. New Jersey initially allowed suffrage for unmarried, property-owning women, but by 1807, no women in the United States had the right to vote.

Women worked for over seventy years to change that. In 1848, at the Seneca Falls women's rights convention, reformers first resolved to seek the right to vote. Then, as now, the question of whose voices should be heard in government was fraught with controversy.

The suffrage movement experienced triumphs and setbacks along the way, but women's voting rights began to spread slowly from west to east. Although Wyoming's territorial legislature was the first to pass a women's suffrage law, Utah Territory quickly followed in 1870. Due to the timing of elections, Utah women became the first in the nation's history to vote with equal suffrage rights open to all female citizens.

The idea of women voting in Utah shocked the nation. Utah seemed the most unlikely setting for advancements in women's rights. At that time, the territory was populated almost exclusively by members of The Church of Jesus Christ of Latter-day Saints, who believed in the practice of plural marriage, or polygamy. Most Americans condemned polygamy as immoral and degrading to women. In 1856, the first Republican Party platform classified polygamy and slavery as "twin relics of barbarism" that should be eliminated. A few years later, Congress passed the anti-polygamy Morrill Act, but this law was essentially unenforced. Utah's involvement within the suffrage movement stirred nationwide controversy and debate about women's rights, religious freedom, and citizenship.

When Utah's territorial legislature unanimously passed the women's suffrage law in 1870, many anti-polygamists and suffragists hoped that Utah women would use their new political power to eradicate polygamy. As it became clear that Latter-day Saint women's votes were in fact supporting Church leaders, federal lawmakers began efforts to revoke Utah women's suffrage. Confronted with these congressional attacks, Latter-day Saint women in Utah stepped onto the national political stage to speak in defense of their religious and political rights. They viewed themselves as cooperative partners with male Church members in working for the good of their community. Using the organizational network and leadership skills they gained through the Church's Relief Society, Latter-day Saint women demonstrated that they could be an influential political force. They held protest meetings, petitioned Congress, lobbied for support from national suffrage leaders, and published a pro-suffrage newspaper, the *Woman's Exponent*.

Of course, not all Utah women felt the same way. Several prominent women who were not Latter-day Saints vehemently opposed polygamy and worked to generate support for anti-polygamy legislation. Although many of these women were committed suffragists, they opposed suffrage for Latter-day Saint women, and a few even opposed all Utah women's voting rights as long as polygamy continued. In 1887, Congress revoked all Utah women's suffrage as part of a new anti-polygamy law.

Outraged that the federal government took their suffrage rights away after seventeen years of voting, most Utah suffragists mobilized and worked together to regain the ballot. They organized the Utah Woman Suffrage Association to ensure that women's voting rights would be included in the state constitution once Utah became a state. Suffragists' efforts succeeded when Utah entered the Union in 1896 as the third suffrage state.

Many Utah women continued to work toward a national constitutional amendment for women's suffrage. They signed petitions, testified before Congress, spoke at national conventions, and a few even picketed the White House. Their commitment to advocating for the rights of women reveals that they were mindful, strategic, and dedicated participants in the cause.

The Nineteenth Amendment prohibited gender-based voting restrictions when it became national law in 1920, and Utah women celebrated this historic step toward women's equality. It marked the achievement of millions of suffragists spanning multiple generations, and although there was still much to do to ensure that all people's voices could be heard in government, Utah women had helped pave the way.

This book honors the many "thinking women" in Utah's history who sought a wider sphere of influence and worked for women's voting rights both locally and nationally. These women made the larger story of suffrage possible, and their legacy inspires us to carry their work forward.

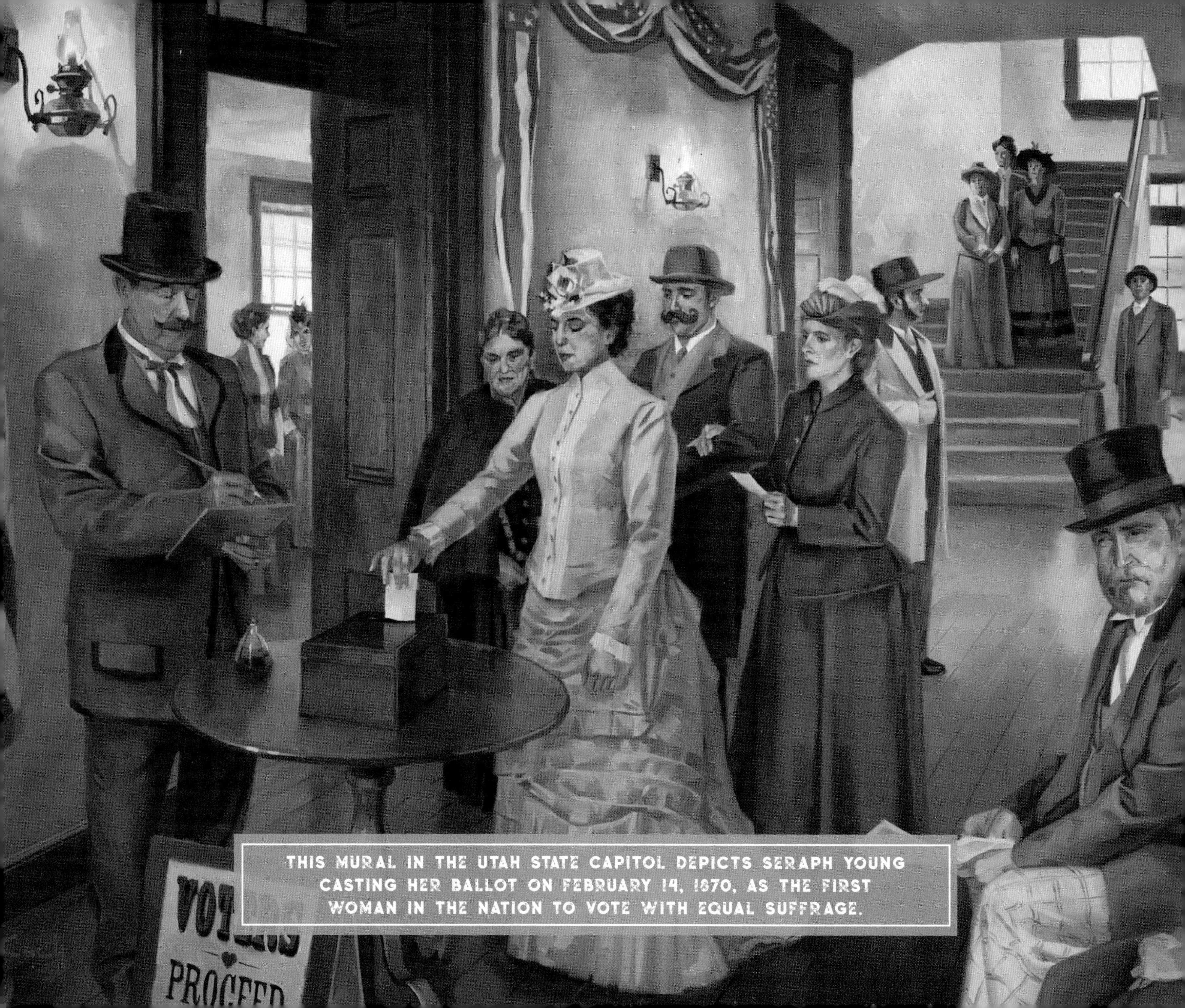

THIS MURAL IN THE UTAH STATE CAPITOL DEPICTS SERAPH YOUNG CASTING HER BALLOT ON FEBRUARY 14, 1870, AS THE FIRST WOMAN IN THE NATION TO VOTE WITH EQUAL SUFFRAGE.

I GAINING THE VOTE 1867–1870

UTAH WOMEN MADE HISTORY IN 1870 BY VOTING IN THE NATION'S first election held with equal suffrage for women. The end of the Civil War in 1865 and the arrival of the transcontinental railroad in 1869 had eroded Utah Territory's isolation and accelerated its entry onto the national political stage. In 1870, the vast majority of Utah residents belonged to The Church of Jesus Christ of Latter-day Saints. With the influx of new arrivals who were not Latter-day Saints, local tensions escalated between Utah's religious and political majority and the small but increasing minority. These shifting dynamics helped turn attention to women's suffrage in Utah.

As the post–Civil War nation considered granting suffrage to African American men, intensifying debates over women's voting rights divided the national suffrage movement and became entangled with the rising national anti-polygamy movement. Arguments both supporting and opposing women's suffrage in Utah were linked with the controversial practice of polygamy from the start. Following some eastern suffragists' suggestions to experiment with women's suffrage in the territories, some anti-polygamists proposed empowering Utah's women with voting rights as a way to end polygamy. Others warned that women's suffrage in Utah would only increase the political power of the Church. In Utah, former Latter-day Saint women in the dissident "New Movement" established the first connections with national suffrage leaders.

Neither pawns nor militants, many of Utah's leading Latter-day Saint women responded to these dynamics by seeking a stronger political voice to join with Latter-day Saint men in defending their religious practices. In January 1870, a large gathering of Relief Society women adopted a resolution calling for the right to vote. The next week, they led a mass meeting of several thousand women to protest a congressional anti-polygamy bill. Just weeks after that powerful display, the Utah Territorial Legislature unanimously passed a law giving voting rights to Utah's female citizens. Wyoming Territory had passed women's suffrage in December 1869, so Utah Territory became the second in the nation to extend equal voting rights to women. Since Utah held both municipal and general elections before Wyoming's next election, Utah women gained the distinction of voting first. They exercised their newfound right in large numbers.

THE FEMALE SUFFRAGE QUESTION.

The New York *Times says*:—

"Female suffrage might perhaps be tried with novel effect in the Territory of Utah—the State of Deseret. There the 'better half' of humanity is in such strong numerical majority that even if all the other half should vote the other way they would carry the election. Perhaps it would result in casting out polygamy and Mormonism in general. And, to prevent woman voters from being under the control of their husbands, they should be allowed to employ 'sealed' ballots. Here would be a capital field for woman suffrage to make a start, and we presume nobody would object to the experiment. Why will not Messrs. Train, Anthony, Stone, and other gentlemen engaged in the cause, turn their attention to this promising field?"

The people of Utah are not afraid of the consequences of giving the women of the Territory the right to vote. In an ecclesiastical capacity they have, from the first organization of the Church of Jesus Christ of Latter-day Saints, had the right. Semi-annually they, with the male members of the Church, vote in General Conference upon all questions which come before the members of the Church for their action. At each Conference the principal authorities of the Church are presented to the people, males and females, for them to vote for or not as they please. In this manner women have for years exercised the right of suffrage in this Territory.

[Special to the Deseret Evening News.]

By Telegraph.

1867

The *New York Times* was the first to publicly advocate enfranchising the women of Utah Territory. A December 17 editorial noted that in Utah, women had "such strong numerical majority" that "they would carry the election" and perhaps even cast out polygamy.

1868

The *Deseret Evening News*, Utah's first newspaper and the official organ of The Church of Jesus Christ of Latter-day Saints, reprinted the *New York Times* editorial on January 9. The paper included its own editorial in favor of women's suffrage in Utah, pointing to Utah's history of women voting in Church matters.

In April, Brigham Young commissioned Eliza R. Snow to help reorganize the Relief Society, the Latter-day Saint women's organization, which had been disbanded for over a decade. Snow traveled throughout Utah Territory to establish local Relief Societies.

"[UTAH] WOULD BE A CAPITAL FIELD FOR WOMAN SUFFRAGE TO MAKE A START, AND WE PRESUME NOBODY WOULD OBJECT TO THE EXPERIMENT."

—NEW YORK TIMES, *DECEMBER 17, 1867*

"THE PEOPLE OF UTAH ARE NOT AFRAID OF THE CONSEQUENCES OF GIVING THE WOMEN OF THE TERRITORY THE RIGHT TO VOTE. IN AN ECCLESIASTICAL CAPACITY . . . WOMEN HAVE FOR YEARS EXERCISED THE RIGHT OF SUFFRAGE IN THIS TERRITORY."

—DESERET EVENING NEWS, *JANUARY 9, 1868*

1866 | 1867 | 1868

"TO ME IT WAS QUITE A MISSION, AND I TOOK MUCH PLEASURE IN ITS PERFORMANCE. I FELT QUITE HONORED AND MUCH AT HOME IN MY ASSOCIATIONS WITH THE BISHOPS, AND THEY APPRECIATED MY ASSISTANCE."
—ELIZA R. SNOW

A Record -

Record of the Smithfield Branch of Zions Female Relief Society

Commenced May 12th 1868.
Closed July 15th 1878
Elizabeth T. Moonhead Prest
Adaline Barber, Mary Ainscough } Councillors
Louisa L. Greene Secty
Margaret Story Treasurer

"WE HAVE MANY TALENTED WOMEN AMONG US, AND WE WISH THEIR HELP IN THIS MATTER [OF CARING FOR THE POOR]. . . . YOU WILL FIND THAT THE SISTERS WILL BE THE MAINSPRING OF THE MOVEMENT. WE RECOMMEND THESE FEMALE RELIEF SOCIETIES TO BE ORGANIZED IMMEDIATELY."
—BRIGHAM YOUNG

The Relief Society was originally established in 1842 in Nauvoo, Illinois, to organize the charitable and spiritual efforts of women in The Church of Jesus Christ of Latter-day Saints. Its reorganization by Brigham Young in 1867 initiated a new era in which Relief Society women worked collectively to improve their communities. More than one hundred local Relief Society chapters were in operation by 1870, and nearly three hundred were active by 1880, offering a new and effective structure for quickly mobilizing the majority of women in Utah. Relief Society meetings gave women opportunities for public speaking and facilitated a deep sisterhood among them. Additionally, women developed important organizational and economic skills through activities such as large-scale silk production, mercantile cooperatives, and grain-saving efforts. The Relief Society's organizational structure served as the foundation for women's forthcoming suffrage advocacy in Utah Territory.

1869 | 1870 | 1871

1868

The Fourteenth Amendment to the United States Constitution was ratified on July 9, granting citizenship to "all persons born or naturalized in the United States" and providing equal protection under the law. The amendment also added the first mention of gender into the Constitution by referring to the right to vote of "male citizens."

Indiana Representative George Washington Julian introduced a bill in Congress on December 14 to grant suffrage to female citizens in all U.S. territories, including Utah. This was the first legislative attempt at women's suffrage in Utah.

"IF THAT WORD 'MALE' BE INSERTED, IT WILL TAKE US A CENTURY AT LEAST TO GET IT OUT."
—ELIZABETH CADY STANTON

1869

A February article published in *Utah Magazine* supported women's equal rights and advocated extending suffrage to women in Utah. The periodical was edited by leaders of the dissident New Movement, including William Godbe.

Congress passed the Fifteenth Amendment on February 26, granting African American men the right to vote. The amendment was ratified the next year on February 3, 1870, just a week before Utah extended suffrage to women.

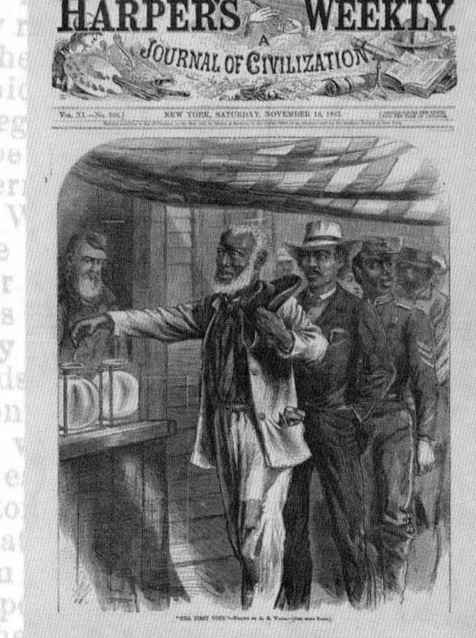

1866 1867 1868

New York suffragist James K. Hamilton Wilcox of the Universal Franchise Association testified before the U.S. House of Representatives Territories Committee on February 27 in favor of Representative Julian's suffrage bill. Wilcox hoped that Utah women would end polygamy if they could vote, but the bill was rejected in committee.

Articles in the *New York Times* on March 5 and 17 reversed the paper's prior suggestion to enfranchise Utah's women, reflecting popular concerns about Latter-day Saint women voting as a bloc to support Church leadership. The paper stated that "the downfall of polygamy is too important to be imperiled by experiments in woman suffrage."

On March 15, Representative Julian introduced "A Bill to Discourage Polygamy in Utah," which simply proposed giving Utah women the right to vote. In endorsing the bill on March 24, the *Deseret Evening News* called itself an "earnest advocate for Women's Rights." The bill eventually died in committee.

William Henry Hooper, Utah's territorial delegate to Congress, surprised anti-polygamists by expressing his support for Representative Julian's bill. He discussed the idea of women's suffrage with Brigham Young and was later instrumental in the passage of Utah's suffrage bill.

FEMALE SUFFRAGE.
Wouldn't it put just a little too much power into the hands of Brigham Young, and his tribe?

"THE PLAN OF GIVING OUR LADIES THE RIGHT OF SUFFRAGE IS, IN OUR OPINION, A MOST EXCELLENT ONE. UTAH IS GIVING EXAMPLES TO THE WORLD ON MANY POINTS, AND IF THE WISH IS TO TRY THE EXPERIMENT OF GIVING FEMALES THE RIGHT TO VOTE IN THE REPUBLIC, WE KNOW OF NO PLACE WHERE THE EXPERIMENT CAN BE SO SAFELY TRIED AS IN THIS TERRITORY. OUR LADIES CAN PROVE TO THE WORLD THAT . . . WOMEN CAN BE ENFRANCHISED WITHOUT RUNNING WILD OR BECOMING UNSEXED."

—DESERET EVENING NEWS, *MARCH 24, 1869*

[Printer's No., 29.

41st CONGRESS, 1st Session.

H. R. 64.

IN THE HOUSE OF REPRESENTATIVES.

March 15, 1869.

Read twice, referred to the Committee on Territories, and ordered to be printed.

Mr. Julian, on leave, introduced the following bill:

A BILL

To discourage polygamy in Utah by granting the right of suffrage to the women of that Territory.

1 *Be it enacted by the Senate and House of Representa-*
2 *tives of the United States of America in Congress assembled,*
3 That from and after the passage of this act the right of
4 suffrage in the Territory of Utah shall belong to, and may be
5 exercised by, the people thereof, without any distinction or
6 discrimination whatever founded on sex.

1869 | 1870 | 1871

The Suffrage Movement Splits: Abolitionists and early women's rights advocates formed the American Equal Rights Association in New York on May 10, 1866, with Lucretia Mott as president. Its purpose was "to secure Equal Rights to all American citizens, especially the right of suffrage, irrespective of race, color or sex." However, at the annual meeting in 1869, deep divisions emerged over the pending Fifteenth Amendment to the U.S. Constitution. Society members disagreed about whether to demand that the amendment extend voting rights to women in addition to black men. This led to a split in the equal rights movement and the formation of two competing women's suffrage organizations: the National Woman Suffrage Association (NWSA), which refused to support an amendment that prohibited race-based voting restrictions but said nothing about gender; and the American Woman Suffrage Association (AWSA), which supported an amendment enfranchising black men as a step toward securing women's suffrage. The divisions between these organizations over strategy and priorities directly impacted Utah's place within the suffrage movement.

1869

The first transcontinental railroad was completed on May 10 with a ceremonial golden spike driven at Promontory Summit, Utah. The influx of railroad workers and new arrivals heightened political and economic tensions in Utah and drew increased national attention to the territory.

In the wake of disputes among suffragists over the Fifteenth Amendment, Susan B. Anthony and Elizabeth Cady Stanton formed the National Woman Suffrage Association (NWSA) on May 15 to pursue a more focused approach to obtain a women's suffrage amendment and other women's rights.

1866 1867 1868

George Q. Cannon, editor of the *Deseret News* and Second Counselor to Brigham Young, published an editorial on May 26 supporting Utah women's involvement in reform efforts.

The 15th Ward Relief Society Hall, dedicated in Salt Lake City on August 5, was a testament to women's economic cooperation. The first floor held a cooperative store financed, owned, and operated by women, and the story above was a meeting space. The hall was the site of several key meetings related to suffrage and was the first of several Relief Society halls in Utah.

In response to the organization of the NWSA earlier that year, Lucy Stone, Henry Blackwell, and Julia Ward Howe formed the American Woman Suffrage Association (AWSA) in November. This more conservative organization was less willing to include Latter-day Saint suffragists in its ranks for fear of losing respectability, so the majority of Utah suffragists ultimately affiliated with NWSA instead.

"WITH WOMEN TO AID IN THE GREAT CAUSE OF REFORM, WHAT WONDERFUL CHANGES CAN BE EFFECTED! WITHOUT HER AID HOW SLOW THE PROGRESS! GIVE HER RESPONSIBILITY, AND SHE WILL PROVE THAT SHE IS CAPABLE OF GREAT THINGS; BUT DEPRIVE HER OF OPPORTUNITIES, MAKE A DOLL OF HER, LEAVE HER NOTHING TO OCCUPY HER MIND . . . AND HER INFLUENCE IS LOST. . . . SUCH WOMEN MAY ANSWER IN OTHER PLACES AND AMONG OTHER PEOPLE; BUT THEY WOULD BE OUT OF PLACE HERE."

—DESERET NEWS, *MAY 26, 1869*

1869 1870 1871

Charlotte Cobb Godbe Kirby (1837–1908) fulfilled her promise to her dying mother that she would continue her suffrage work. An ambitious feminist, Charlotte Cobb Godbe (later Kirby) was the first Utah woman to publicly advocate for women's equal rights on a national platform. She was also the first Utah woman to hold office in a national suffrage organization, serving on the NWSA education committee beginning in May 1871. Godbe's associations with eastern suffrage leaders and her letters to eastern suffrage newspapers helped the women of Utah build ties to the national suffrage movement. She lectured on suffrage and temperance in cities such as Boston and Washington, D.C., and also in Utah towns including Logan and American Fork. Godbe straddled the complicated dynamics associated with the suffrage movement in Utah. A plural wife of the dissident New Movement leader William Godbe, she eventually divorced and rejected plural marriage. She was not accepted by Latter-day Saint suffragists as one of their own, although she corresponded with Church leaders and defended Latter-day Saint women's voting rights nationally.

1869

Augusta Adams Cobb Young, one of Brigham Young's plural wives, traveled with her daughter Charlotte Cobb to visit eastern suffragists such as Lucy Stone in Boston. Augusta and Charlotte were the first women from Utah to establish relationships with national suffrage leaders.

On December 10, Wyoming Territory made history as the first state or territory in the nation to grant equal suffrage to women citizens. The territorial law also allowed women to run for public office. Wyoming's population at the time included almost 1,500 female citizens of voting age.

FEMALE SUFFRAGE.

CHAPTER 31.

AN ACT TO GRANT TO THE WOMEN OF WYOMING TERRITORY THE RIGHT OF SUFFRAGE, AND TO HOLD OFFICE.

Be it enacted by the Council and House of Representatives of the Territory of Wyoming:

SEC. 1. That every woman of the age of twenty-one years, residing in this territory, may, at every election to be holden under the laws thereof, cast her vote. And her rights to the elective franchise and to hold office shall be the same under the election laws of the territory, as those of electors. *Women's rights.*

SEC. 2. This act shall take effect and be in force from and after its passage.

Approved, December 10th 1869.

1866 1867 1868

"WE DEMAND OF THE GOV. THE RIGHT OF FRANCHISE"

—*BATHSHEBA W. SMITH*

"WE ARE NOT INFERIOR TO THE LADIES OF THE WORLD AND WE DO NOT WANT TO APPEAR SO. . . . THE LADIES OF UTAH [HAVE] TOO LONG REMAINED SILENT WHILE THEY WERE BEING SO FALSELY REPRESENTED TO THE WORLD."

—*ELIZA R. SNOW*

1870

On January 6, Latter-day Saint women met in the 15th Ward Relief Society Hall with President Sarah M. Kimball as chair. They organized a protest against proposed federal legislation that would deny citizenship and voting rights to all men practicing polygamy. Recognizing the power of a political voice, the women voted to demand the "right of franchise" and to send their own representatives to Washington, D.C. This marked Utah women's first organized efforts seeking their own right to vote.

1869 | 1870 | 1871

1870 *In a dramatic entry into collective political activity, over five thousand Latter-day Saint women gathered for a "Great Indignation Meeting" to protest anti-polygamy legislation on January 13 in Salt Lake City's Old Tabernacle. In their speeches, the women demonstrated that they were intelligent and articulate defenders of their religious rights.*

1866 1867 1868

"LET [REP. CULLOM] LEARN THAT [THE WOMEN OF UTAH] ARE *ONE* HEART, HAND AND BRAIN, WITH THE BROTHERHOOD OF UTAH."

—HANNAH T. KING

"WE ARE NOT HERE TO ADVOCATE WOMAN'S RIGHTS, BUT MAN'S RIGHTS. THE BILL IN QUESTION WOULD . . . DEPRIVE OUR FATHERS, HUSBANDS AND BROTHERS OF ENJOYING THE PRIVILEGES BEQUEATHED TO CITIZENS OF THE UNITED STATES."

—SARAH M. KIMBALL

"WERE WE THE STUPID, DEGRADED, HEARTBROKEN BEINGS THAT WE HAVE BEEN REPRESENTED, SILENCE MIGHT BETTER BECOME US; BUT, AS WOMEN OF GOD, . . . WE NOT ONLY SPEAK BECAUSE WE HAVE THE RIGHT, BUT JUSTICE AND HUMANITY DEMAND THAT WE SHOULD."

—ELIZA R. SNOW

"LET THE UNITED VOICE OF THIS ASSEMBLY GIVE THE LIE TO THE POPULAR CLAMOUR THAT THE WOMEN OF UTAH ARE OPPRESSED AND HELD IN BONDAGE."

—HARRIET COOK YOUNG

"I LONG TO SEE THE WOMEN OF UTAH RISE AND EXPRESS THEMSELVES CONCERNING THEIR RIGHTS."

—ELEANOR M. PRATT

"IT WILL NOT BE DENIED THAT THE MORMON WOMEN HAVE BOTH BRAINS AND TONGUES. SOME OF THE SPEECHES GIVE EVIDENCE THAT IN GENERAL KNOWLEDGE, IN LOGIC, AND IN RHETORIC THE SO-CALLED DEGRADED LADIES OF MORMONDOM ARE QUITE EQUAL TO THE WOMEN'S RIGHTS WOMEN OF THE EAST."

—NEW YORK HERALD, *JANUARY 23, 1870*

1869 1870 1871

1870

At the NWSA suffrage convention in Washington, D.C., on January 18, Susan B. Anthony proposed a resolution that Congress "enfranchise the women of Utah, as the one safe, sure and swift means to abolish the polygamy of that Territory." The convention unanimously adopted her resolution.

The Utah Territorial Legislature, composed entirely of Latter-day Saint men, began considering a women's suffrage bill on January 27. "A number of ladies, for whom seats had been provided, graced the debate with their presence, and listened with much interest."

A handwritten draft of "An Act In Relation to Suffrage" reveals some of the details that the legislature initially considered, including a lower voting age of eighteen for both women and men. The ultimate bill kept the voting age at twenty-one for all voters.

An act in relation to Suffrage

An Act in relation to Suffrage.

Sec 1 Be it enacted by the Governor and Legislative Assembly of the Territory of Utah: That all male citizens of the United States over the age of 21 years ~~eighteen years, either male and female,~~ over the age of 18 years who have actually resided in the precinct one year next preceding any election, shall be entitled to vote for Territorial County precinct and Municipal officers in the precinct where they reside.

Sec 2. ~~And be it further enacted, that so much of~~ an act, entitled, "An act prescribing certain qualifications necessary to enable a person to be eligible to hold office, vote or serve as a juror, approved Jan 21st 1859" as conflicts with this act is hereby repealed -

1866 | 1867 | 1868

Led by Speaker Orson Pratt, the Utah House of Representatives unanimously passed a bill extending the right to vote to women citizens on February 5. The bill was then sent to the Legislative Council (Utah's territorial senate) for consideration, where it was debated and amendments were proposed.

A *Deseret News* editorial on February 8 voiced strong support for Utah women's suffrage and declared the paper a "decided advocate of the rights of women."

On February 9, Utahns who were opposed to the political power of the Latter-day Saint Church joined forces to form the Liberal Party, the territory's first political party. Latter-day Saints quickly responded by forming the People's Party.

THE EVENING NEWS.

GEORGE Q. CANNON,

EDITOR AND PUBLISHER.

Tuesday, - - February 8, 1870.

FEMALE SUFFRAGE IN UTAH.

THE female suffrage question is now fairly before the nation; its advocates are as earnest in their labors as if the salvation of the world depended upon their success, and the triumph of the movement, we believe, is only a question of time. The agitation of the question has reached the Rocky Mountains. In our neighboring Territory, Wyoming, the cause has triumphed; in Colorado the ladies are petitioning to have female suffrage legalized there. But success by piecemeal will not satisfy those who are acknowledged as the national leaders of the movement; nothing short of an amendment to Constitution of the United States to this effect will do for them, and this is now being eagerly sought; and as the Congressmen are noted, among other things, for their gallantry and their susceptibility to female charms, the adoption of such an amendment is not at all improbable.

We believe in the right of suffrage being enjoyed by all who can exercise it intelligently; but our lawmakers, in conferring this great power upon the recently emancipated black race, do not seem to regard intelligence as an indispensible pre-requisite; and we think the suffrage might be conferred with much greater propriety upon intelligent white women than upon ignorant blacks.

The idea of female suffrage is regarded by many as peculiar to and having originated in these last days; but history tells us that a similar movement existed in ancient Greece when that nation was in the meridian of her splendor. If the right of suffrage was granted to the ladies then it certainly did not bring about the reforms considered necessary to preserve that nation from decadence, and whether it would in this is extrem-

...

but if this bill passes they will be their equals in that respect too. We are satisfied that the result will be exactly opposite to what our enemies anticipate. On the plural marriage question we are as firmly convinced as we are of our own existence that were its continuance or abolition put to the vote of the female portion of our population to-day it would be sustained by a nine-tenths majority; and upon this score, which has enlisted the mock sympathy of so many, no disadvantage to Zion's cause will ensue. In every other it cannot but result also in good. We have many *friends* around whose constant effort is to out-vote the "Mormons" at their municipal elections so that the discordant elements so overwhelmingly developed in municipal rule everywhere but in Zion might be introduced here. Many of our cotemporaries boast that this consummation will soon be brought about now that direct rail communication exists between the cities of Utah and the East and West. We do not anticipate such a result; nevertheless the hopes of our enemies in this respect may be realized. We do not believe, however, that the existence of our most cherished institutions depends on such a frail tenure as the possession of power by the female members of the Church to vote them down. If such be the case, we believe the ladies should have the power to exercise their agency, hence we desire to see the matter tested; and we hope that the bill passed by the House of Representatives of the Territorial Legislature, on Saturday, will be passed by the Council, believing that the result will be an additional proof to the world, that even with this power in their hands the ladies of Utah will remain true to their integrity, and then, as now, will sustain the priesthood, whether acting in a religious or civil capacity, in promoting the cause of Zion and the behests of Heaven.

[SPECIAL TO THE DESERET NEWS.]

By Telegraph.

AFTERNOON DISPATCHES.

1870

Both the House and the Legislative Council of the Utah territorial legislature unanimously passed an amended women's suffrage bill on February 10. This bill conferred equal voting rights, but not the right to hold office, upon the women citizens of Utah.

Acting Governor Stephen A. Mann reluctantly signed the suffrage bill into law on February 12. Mann was convinced to do so by the territorial legislature's unanimous approval despite his own "very grave and serious doubts of the wisdom" of the bill. The groundbreaking legislation officially granted voting rights to approximately 18,000 women in Utah Territory, swelling the ranks of eligible female voters in the nation by twelve times.

AN ACT

Conferring upon Women the Elective Franchise.

[Approved February 12, 1870.]

SEC. 1. *Be it enacted by the Governor and Legislative Assembly of the Territory of Utah:* That every woman of the age of twenty-one years who has resided in this Territory six months next preceding any general or special election, born or naturalized in the United States, or who is the wife, widow or the daughter of a native-born or naturalized citizen of the United States, shall be entitled to vote at any election in this Territory.

SEC. 2. All laws or parts of laws conflicting with this Act are hereby repealed.

1866 | 1867 | 1868

On February 14, two days after women's suffrage became law in Utah, about twenty-five women voted in a Salt Lake City municipal election at Council Hall. This marked the first time in the nation that women voted with equal suffrage.

Seraph Young, a schoolteacher and the grand-niece of Brigham Young, was the first woman to cast a ballot. Despite the historic significance of her vote, little is known of her personal life—a reminder that ordinary people make history.

Mary Gibbs-Bigelow and Lucy Bigelow Young were among the other women who voted that day amid stump speeches and music from the 10th Ward brass band.

1869 1870 1871

A ladies' reform or "retrenchment" meeting held in the 15th Ward Relief Society Hall on February 19, 1870, quickly turned to a discussion of the newly granted right of suffrage. When Sarah M. Kimball asked for a showing of those who supported women's rights, many of the women "manifested their approval." The women expressed a range of views on the subject of women's rights, but all were committed to fulfilling their new responsibility as voters.

"I HAVE WAITED PATIENTLY A LONG TIME, AND NOW THAT WE WERE GRANTED THE RIGHT OF SUFFRAGE, I CAN OPENLY DECLARE MYSELF A WOMAN'S RIGHTS WOMAN."
—*SARAH M. KIMBALL*

"WE MUST ACT IN WISDOM, AND NOT GO TOO FAST, . . . AND NOT RUN HEADLONG AND ABUSE THE PRIVILEGE."
—*PHOEBE WOODRUFF*

"I AM GLAD TO SEE OUR DAUGHTERS ELEVATED WITH MAN."
—*PRESENDIA KIMBALL*

"WOMAN'S RIGHTS HAVE BEEN SPOKEN OF. I HAVE NEVER HAD ANY DESIRE FOR MORE RIGHTS THAN I HAVE. I HAVE ALWAYS CONSIDERED THESE THINGS BENEATH THE SPHERE OF WOMAN. BUT AS THINGS PROGRESS I FEEL IT IS RIGHT THAT WE SHOULD VOTE."
—*MARGARET T. SMOOT*

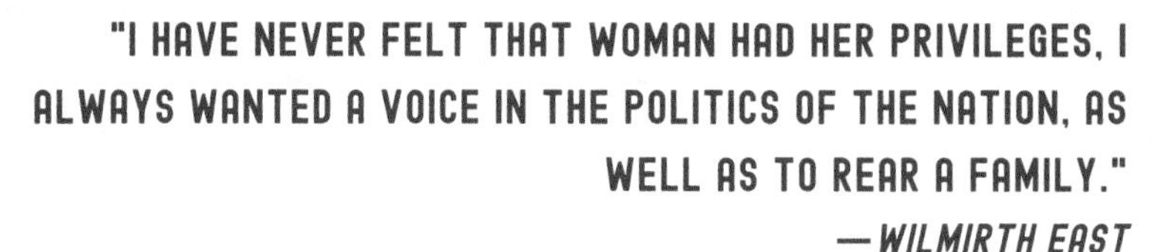

"I HAVE NEVER FELT THAT WOMAN HAD HER PRIVILEGES, I ALWAYS WANTED A VOICE IN THE POLITICS OF THE NATION, AS WELL AS TO REAR A FAMILY."
—*WILMIRTH EAST*

1866 | 1867 | 1868

1870

Bathsheba W. Smith was appointed to travel to southern Utah to preach about women's reforms and "woman's rights if she wished."

At the February 19 meeting, Relief Society President Eliza R. Snow moved that a committee draft an "expression of gratitude" to Acting Governor Mann. He responded with a letter "expressing the confident hope, that the ladies of this Territory will so exercise the right conferred as to approve the wisdom of the legislation."

By March, an estimated 25,000 women had participated in mass meetings throughout the territory to protest proposed anti-polygamy legislation. The *Deseret News* asserted that these "women's rights meetings . . . of the ladies of Utah are deserving of consideration by all."

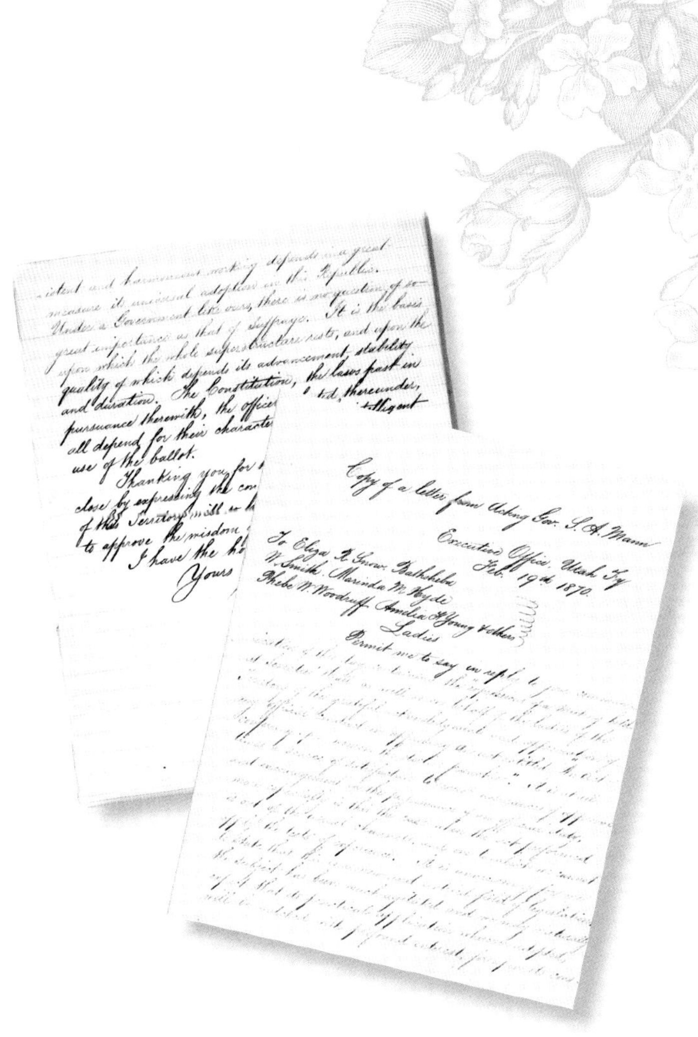

…istent and harmonious working depends in a great measure its universal adoption in this Republic.
Under a Government like ours, there is no question of so great importance as that of Suffrage. It is the basis upon which the whole superstructure rests, and upon the quality of which depends its advancement, stability and duration. The Constitution, the laws past in pursuance therewith, the officer[illegible] …ted, thereunder, all depend for their characte[illegible] …telligent use of the ballot.
Thanking you for [illegible] close by expressing the con[illegible] of this Territory will so [illegible] to approve the wisdom [illegible]
I have the ho[illegible]
Yours [illegible]

Copy of a letter from Acting Gov. S. A. Mann
Executive Office, Utah Ty
Feb. 19th 1870.
To Eliza R. Snow. Bathsheba W. Smith. Marinda M. Hyde Phebe W. Woodruff. Amelia F. Young & others
Ladies
Permit me to say in reply to [illegible]

THE ELECTION.—Quietness was the order of the day yesterday. Everybody appeared to have read the HERALD attentively on Sunday, and to have taken the sensible advice given therein. The 10th ward brass band, Captain Beesely's martial band and Olsen's string band were out, and discoursed some excellent music. Brief visits to the polling places gave us to understand that a large number of ladies were exercising the lately granted right of the franchise. And though there was considerable good-humored chaffing, the utmost respect was shown by all to the ladies, for whom a separate entrance to the place of voting was provided.

This election will be memorable in the history of the Territory as the first Territorial election at which women exercised the franchise. There have been municipal elections in different places, since Hon. S. A. Mann attached his signature to the act conferring the suffrage upon them; but this is the first time the women of Utah have had an opportunity to express by their ballots their sentiments on a leading public question—whether they, the parties most deeply concerned, would sustain polygamy or repudiate it; for this question has been lugged into the election and forced upon the voters by a few who believed in a fight, no matter how great the fizzle they would make. The result of the polling, when known, will show that the women emphatically sustain their husbands, fathers and brothers, their domestic institutions, their hearths and homes every time, before a few dissension-breeding "carpet-baggers"—there, the word's out [illegible] we can't help it. And, now, we [illegible]: Hurrah for the women of

1870

Sarah M. Kimball and other female leaders used the Relief Society organization to encourage voting and to teach civics, history, political science, and parliamentary law classes.

Thousands of Utah women cast their votes in the territory's general election on August 1. It was the first general election in the nation in which women voted with equal suffrage.

George A. Smith, First Counselor to Brigham Young, wrote that he "had the pleasure of staying at home and taking care of the children while my wives went to the election."

On September 6, Wyoming women voted in the first election held in the territory since the historic suffrage bill was passed the previous year.

"THIS ELECTION WILL BE MEMORABLE IN THE HISTORY OF THE TERRITORY AS THE FIRST TERRITORIAL ELECTION AT WHICH WOMEN EXERCISED THE FRANCHISE. . . . HURRAH FOR THE WOMEN OF UTAH."

—SALT LAKE HERALD-REPUBLICAN, *AUGUST 2, 1870*

1866 1867 1868

Charlotte Cobb Godbe returned to Boston to visit with eastern suffragists such as Mary Livermore, the coeditor of the AWSA's Woman's Journal. Godbe spoke at a women's suffrage meeting in Providence, Rhode Island, where she was the first American woman with voting rights to address eastern suffragists.

A "Mrs. Godbe" was listed "among the distinguished guests" in New York City on October 20 at the twentieth anniversary celebration of the first national women's rights convention. In her keynote remarks, New York suffragist Paulina W. Davis dismissed Utah suffrage as "of less account, because the women there are more under a hierarchy than elsewhere, and as yet vote only as directed."

Charlotte Cobb Godbe published a December 15 article in *The Revolution*, Elizabeth Cady Stanton and Susan B. Anthony's suffrage newspaper, in which she predicted that Utah women would use the vote to end polygamy. Such representations encouraged national suffragists, who struggled to reconcile Utah's polygamy with its suffrage.

"UTAH IS A LAND OF MARVELS. SHE GIVES US, FIRST, POLYGAMY, WHICH SEEMS TO BE AN OUTRAGE AGAINST 'WOMAN'S RIGHTS,' AND THEN OFFERS THE NATION A 'FEMALE SUFFRAGE BILL,' AT THIS TIME IN FULL FORCE WITHIN HER OWN BORDERS. WAS THERE EVER A GREATER ANOMALY KNOWN IN THE HISTORY OF SOCIETY?"
—PHRENOLOGICAL JOURNAL, *NOVEMBER 1870*

"AS AMERICA LOOKS TO THE RISING GENERATION FOR HER HOPE, SO, IN THE RISING WOMEN OF UTAH, DO I SEE LEADERS IN PUBLIC THOUGHT."
—*CHARLOTTE COBB GODBE*

EMMELINE B. WELLS, UTAH'S MOST PROMINENT SUFFRAGIST AND EDITOR OF THE *WOMAN'S EXPONENT*, SEATED AT HER DESK IN A PORTRAIT TAKEN JANUARY 14, 1879.

2 INDEPENDENT VOTERS 1871–1881

UTAH WOMEN ATTRACTED NATIONAL ATTENTION AS VOTERS.

In the decade following their first votes, they actively participated in politics and began to establish an official, albeit controversial, relationship with the national suffrage movement. In 1871, NWSA leaders Elizabeth Cady Stanton and Susan B. Anthony visited Utah Territory and personally congratulated Utah's women voters. They spoke in the Old Tabernacle, where the Salt Lake Assembly Hall now stands, and also gave a lecture at the newly-opened Liberal Institute to members of the disaffected New Movement and others not affiliated with the Church. This visit began NWSA's complicated relationship with both polygamous and non-polygamous Utah suffragists.

Utah women's voting did not bring an end to polygamy or weaken Latter-day Saint political power in the territory, as many national observers had hoped. Women on both sides of the polygamy issue began sending petitions to national leaders with the aim of influencing federal legislation regarding Utah Territory. In the face of growing hostility to both polygamy and Utah women's suffrage, Latter-day Saint women portrayed themselves as empowered women and as citizens whose religious and political rights deserved protection. They founded a newspaper, the *Woman's Exponent*, to correct misrepresentations of their religion, education, and interests. Under Louisa Lula Greene Richards and later Emmeline B. Wells, the paper championed women's economic, educational, and political rights, in addition to sharing news about suffrage and the Relief Society. After Wells led Utah women in sending petitions to Washington, D.C., for a women's suffrage constitutional amendment, she and Zina Young Williams represented Utah and spoke at the 1879 NWSA convention in Washington, D.C.

Realizing that the majority of Utah women were using their votes to support the People's Party and Church policies, members of the Liberal Party and the anti-polygamy movement redirected their efforts toward shifting the balance of political power. Failing to invalidate the suffrage law through the Utah court system, they lobbied for federal legislation to end polygamy through several means, including ending Utah women's suffrage. The Ladies' Anti-Polygamy Society of Utah sought to increase public support for anti-polygamy legislation and eventually called for the disenfranchisement of Utah women through its newspaper, the *Anti-Polygamy Standard*. While the larger and more conservative AWSA opposed polygamous women's voting rights, most NWSA leaders continued to defend the principle of women's suffrage despite their personal opposition to polygamy.

Susan B. Anthony (1820-1906), one of America's best known suffragists, spent her life fighting for women's rights. Despite her repudiation of polygamy, Anthony ultimately became a key ally for Utah women within the national suffrage movement. Visiting Utah in 1871 and again in 1895, she formed strong ties with many Utah suffragists, mentoring and supporting them in their pursuit of voting rights. Utah women also helped Anthony campaign for a constitutional women's suffrage amendment. For Anthony's eightieth birthday, the woman-run Utah Silk Commission gave her some black silk entirely produced by Utah women, from which she had a cherished dress made. Later, upon her deathbed, Anthony bequeathed a gold ring to Utah suffragist Emmeline B. Wells.

1871

National suffragist Victoria Woodhull testified before the U.S. House of Representatives Judiciary Committee on January 11, arguing that women already had suffrage rights based on citizenship granted by the Fourteenth Amendment. She was the first woman to address a congressional committee.

New Movement leaders in Utah invited NWSA President Elizabeth Cady Stanton and Vice President Susan B. Anthony to come meet Utah's voting women. Learning of their upcoming visit, Brigham Young invited the preeminent national suffragists to also speak to Latter-day Saints.

Stanton and Anthony arrived in Utah on June 28, and William Godbe met them at the train depot. While in Utah, the suffrage leaders divided their time between Latter-day Saint and New Movement audiences.

1871 1872 1873

Elizabeth Cady Stanton spoke to over a thousand Latter-day Saints in Salt Lake City's Old Tabernacle on June 29 and then gave a five-hour lecture on women's rights to about three hundred women on June 30. While much of Stanton's speech was well-received, her controversial remarks on family planning were too radical for most of her listeners. Susan B. Anthony lectured separately to a small group of Latter-day Saints the evening of June 30.

Susan B. Anthony spoke to Latter-day Saints at the new Tabernacle on July 2. She also gave short remarks that day at the dedication of the Liberal Institute, built by the New Movement as a venue for any who were not Latter-day Saints to gather. Elizabeth Cady Stanton lectured at the Liberal Institute the next evening.

Stanton and Anthony met prominent city and territorial leaders, attended July 4th celebrations in both the new Tabernacle and the Liberal Institute, and dined with suffragists Mary and Annie Godbe, who served as NWSA officers representing Utah in the following years. Stanton and Anthony attended a large meeting of women at the Liberal Institute on July 5 and a small suffrage meeting in Corinne, Utah, on July 7 before leaving Utah by train that evening.

"I WOULD RATHER BE A WOMAN AMONG MORMONS WITH THE BALLOT IN MY HANDS THAN AMONG GENTILES WITHOUT THE BALLOT. IF THERE IS HEREAFTER ANY SLAVERY AMONG THE WOMEN OF UTAH IT IS THEIR OWN FAULT, FOR THEY HOLD THE POWER WITHIN THEIR OWN HANDS TO RID THEMSELVES OF IT. THEIR FIRST THOUGHT SHOULD BE HOW TO USE THE BALLOT FOR THEIR OWN GOOD."

—ELIZABETH CADY STANTON

1874 1875 1876

1871

Speaking at a Pioneer Day celebration in Ogden, Relief Society President Eliza R. Snow differentiated between Latter-day Saint women's goals and those of national women's leaders. Likely responding to Stanton's controversial remarks earlier that month, Snow supported women's suffrage but warned against "strong-minded" women and a "war of sexes" that could arise from the women's rights movement.

Charlotte Godbe was appointed in May as a member of the Executive Committee of NWSA. She was invited to speak to a large audience at Tremont Temple in Boston for a meeting of the New England Suffrage Association.

"IT IS FOR US TO SET THE WORLD AN EXAMPLE OF THE HIGHEST AND MOST PERFECT TYPES OF WOMANHOOD. . . . WHAT INTERESTS HAVE [WOMEN] THAT ARE NOT IN COMMON WITH [MEN'S], AND WHAT HAVE THEY THAT ARE DISCONNECTED WITH OURS? WE KNOW OF NONE."
—*ELIZA R. SNOW*

1870 1871 1872

1872

At a primary election meeting on February 3, Bathsheba W. Smith and Sarah M. Kimball were appointed to the committee to nominate candidates for Salt Lake City offices.

Following its defeat in the February 12 election, the Liberal Party published attacks on Utah women's suffrage in the *Salt Lake Tribune*, arguing that the legislature lacked power to grant women the right to vote and that most women voting were not legally naturalized citizens.

The Liberal Party appealed to Congress in February to limit women's suffrage in Utah, reduce the political power of the majority party, and place Utah elections under federal control.

A constitutional convention adopted a proposed Utah state constitution on March 2 and applied to Congress for statehood for the fifth time. Elizabeth Howard and Wilmirth East had served on the nominating committee to select convention delegates.

"THE LADIES WENT TO THE POLLS YESTERDAY BY THE WAGONLOAD TO VOTE THE THEOCRATIC [PEOPLE'S PARTY] TICKET. NOW WE CLAIM THAT THE WOMAN VOTE IN UTAH IS ILLEGAL, FOR THE SIMPLE REASON THAT THE UTAH LEGISLATURE HAS NO POWER OVER THE QUESTION OF SUFFRAGE."
—SALT LAKE TRIBUNE, *FEBRUARY 13, 1872*

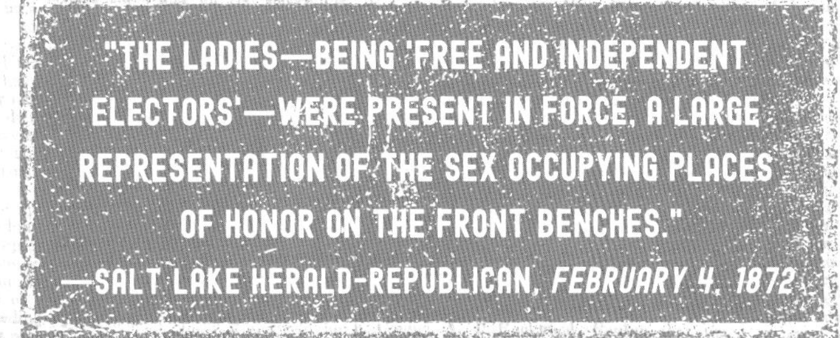

1873 | 1874 | 1875

1872

Led by Cornelia Paddock, Jennie Froiseth, and "Mrs. Godbe," over four hundred Utah women sent a petition to Vice President Schuyler Colfax on March 12 opposing Utah statehood because of polygamy and the political power of Latter-day Saints. This marked the first significant opposition to statehood from within Utah.

Editor Louisa Lula Greene Richards published the first issue of the *Woman's Exponent* in Salt Lake City on June 1. This woman-owned and woman-run newspaper reported on advancements for women across the world, the work of the Relief Society, and other news of interest to Latter-day Saint women.

Phoebe Couzins and Georgiana Snow became the first female attorneys admitted to the Utah Bar on September 25. Couzins, a Missouri resident, later became a NWSA leader and key advocate for Utah's suffrage. Snow, daughter of the Utah Attorney General, paved the way for Utah women to practice law.

In Rochester, New York, Susan B. Anthony led a small group of women to the polls and cast a ballot for Ulysses S. Grant in the national election. Anthony was arrested two weeks later and was eventually fined $100 for illegally voting. She argued that she had a constitutional right to vote as an American citizen and never paid the fine.

"THE WOMEN OF UTAH TODAY OCCUPY A POSITION WHICH ATTRACTS THE ATTENTION OF INTELLIGENT THINKING MEN AND WOMEN EVERYWHERE. . . . WHO ARE SO WELL ABLE TO SPEAK FOR THE WOMEN OF UTAH AS THE WOMEN OF UTAH THEMSELVES? 'IT IS BETTER TO REPRESENT OURSELVES THAN TO BE MISREPRESENTED BY OTHERS!'"

—WOMAN'S EXPONENT, *JUNE 1, 1872*

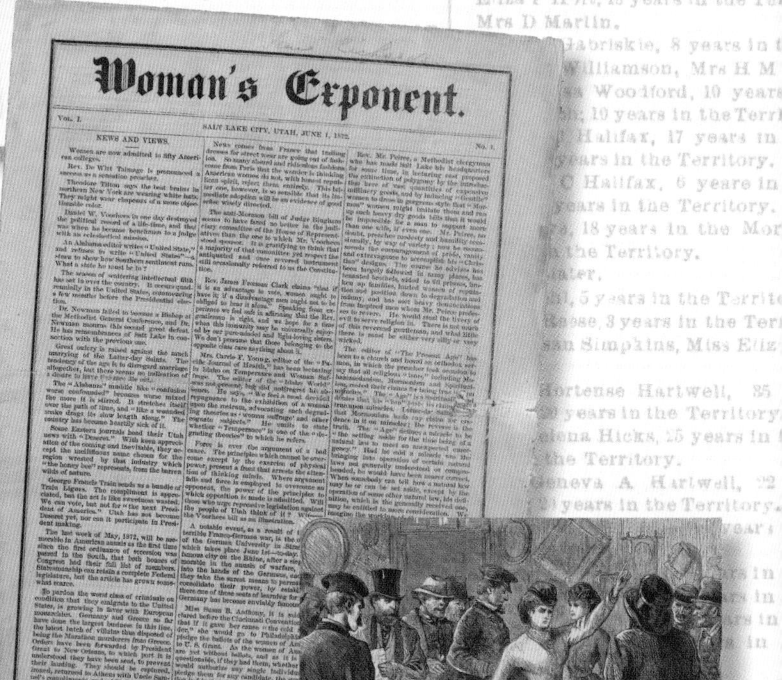

Woman's Exponent.

Vol. 1. SALT LAKE CITY, UTAH, JUNE 1, 1872. No. 1.

NEWS AND VIEWS.

1870 1871 1872

"IT WAS WE, THE PEOPLE, NOT WE, THE WHITE MALE CITIZENS, NOR YET, WE, THE MALE CITIZENS; BUT WE, THE WHOLE PEOPLE, WHO FORMED THE UNION. AND WE FORMED IT . . . NOT TO THE HALF OF OURSELVES AND THE HALF OF OUR POSTERITY, BUT TO THE WHOLE PEOPLE—WOMEN AS WELL AS MEN. AND IT IS A DOWNRIGHT MOCKERY TO TALK TO WOMEN OF THEIR ENJOYMENT OF THE BLESSINGS OF LIBERTY WHILE THEY ARE DENIED THE USE OF THE ONLY MEANS OF SECURING THEM . . .—THE BALLOT."

—*SUSAN B. ANTHONY*

1873

U.S. Senator Frederick Frelinghuysen from New Jersey introduced a "Utah Bill" on February 6, which included clauses to repeal Utah Territory's suffrage act and deprive married women of property rights. It passed the Senate but failed to pass the House, due to the opposition of suffrage organizations throughout the country.

On January 20, Susan B. Anthony, Matilda Gage, and Elizabeth Cady Stanton submitted a petition for equal suffrage to the U.S. House of Representatives Judiciary Committee on behalf of NWSA. The petition included two Utah women among the list of NWSA officers: Sarah Stenhouse as a vice president for Utah and Mary Godbe as a member of the advisory committee.

1874

On June 23, the U.S. Congress passed the Poland Act, which redefined Utah court jurisdictions to replace some local authorities with federal authorities. This was intended to make it easier to convict polygamists under the 1862 Morrill Anti-Bigamy Act.

1873 1874 1875

Emmeline B. Wells (1828–1921), though small in stature, wielded power through her strong will and pen to become Utah's preeminent suffragist. "Aunt Em," as she was known, edited the Woman's Exponent *for almost forty years, making it one of the longest-running women's rights newspapers in the U.S. She spoke at numerous national suffrage conventions, developed a close relationship with Susan B. Anthony and other national and international women's leaders, and met four U.S. presidents through her advocacy work for Utah women. Wells led several women's clubs in Salt Lake City and always sought to build bridges between Latter-day Saint women and women of other beliefs. She served as the fifth President of the Relief Society from 1910 to 1921 and lived to see the Nineteenth Amendment extend women's suffrage throughout the nation in 1920.*

"IN THE CHURCH AND KINGDOM OF GOD, THE INTERESTS OF MEN AND WOMEN ARE THE SAME; MAN HAS NO INTERESTS SEPARATE FROM THAT OF WOMEN, HOWEVER IT MAY BE IN THE OUTSIDE WORLD, OUR INTERESTS ARE ALL UNITED."

—ELIZA R. SNOW

1874

Suffrage leader Victoria Woodhull lectured at the Salt Lake Theatre on July 18. In 1872, Woodhull became the first woman to run for U.S. President, but her controversial stands on free love and other radical issues were often a liability for the suffrage movement. Emmeline B. Wells attended Woodhull's lecture and countered with an article in the *Woman's Exponent* criticizing her for classing all women as "weak, ignorant, vain and silly, slaves to dress, to fashion and to men."

"ONE GREAT FAULT IN 'WOODHULL'S LECTURE' HERE, WAS, THAT WHILE PROFESSING TO HOLD FORTH FOR WOMEN'S RIGHTS, SHE SPOKE SO DISPARAGINGLY OF THE WOMEN THEMSELVES. . . . I BELIEVE IN WOMEN, ESPECIALLY THINKING WOMEN."

—EMMELINE B. WELLS

1874 | 1875 | 1876

"INASMUCH AS IT HAS BEEN FREQUENTLY SAID ABROAD THAT THE WOMEN OF UTAH ACT FROM COMPULSION—WE TESTIFY TO YOU THAT THE GETTING UP OF THIS [PETITION] IS WHOLLY WOMEN'S WORK."

—DESERET NEWS, *FEBRUARY 9, 1876*

1875

On March 29, the U.S. Supreme Court unanimously ruled in the case of *Minor v. Happersett* that citizenship alone did not guarantee women the right to vote. The only recourse left for suffragists was either to convince Congress to pass a national women's suffrage amendment or to lobby state legislatures to pass women's suffrage laws.

Caroline Severance, an abolitionist and founding member of AWSA, visited Salt Lake City on May 30 and met with several prominent Utah women, including both Latter-day Saints and others. The *Woman's Exponent* reported: "We shall remember with much pleasure her words of encouragement."

1876

In January, over 26,000 Latter-day Saint women signed a petition to the U.S. Congress asking for Utah's admission as a state and for the repeal of anti-polygamy legislation.

Women from Utah's Liberal Party submitted a separate petition to Congress to counter the Latter-day Saint women's petition. A delegation of anti-polygamists from Utah met with U.S. President Ulysses S. Grant on January 28 to lobby against polygamy and women's suffrage in Utah.

1877 1878 1879

Belva Lockwood (1830–1917) was a national suffrage leader and the first woman admitted to practice law before the U.S. Supreme Court. Though not a member of the Church of Jesus Christ, she became an important advocate for Utah women's right to vote. While many other national suffragists supported the disenfranchisement of polygamous women, Lockwood vocally opposed what she considered the anti-polygamy movement's broader violations of civil rights. Her defense of Latter-day Saints strained her relationship with other national suffrage leaders, who did not support her when she ran for U.S. president in 1884 as the Equal Rights Party candidate.

1876

The NWSA annual suffrage convention in Washington, D.C., passed a resolution on January 28 denouncing attempts to disenfranchise female voters in Utah as a "gross invasion of a vested right." Belva Lockwood, Sara Andrews Spencer, and Ellen C. Sargent were appointed to "watch over the rights of the women of Utah," and Annie Godbe was selected as the NWSA vice president representing Utah.

In October, Emmeline B. Wells was appointed by Brigham Young to lead the Relief Society's grain storing efforts. She and other Utah women gained valuable organizing and leadership experience as they managed this large-scale gathering of wheat over forty-two years. The grain was used for humanitarian aid and to feed U.S. troops during World War I.

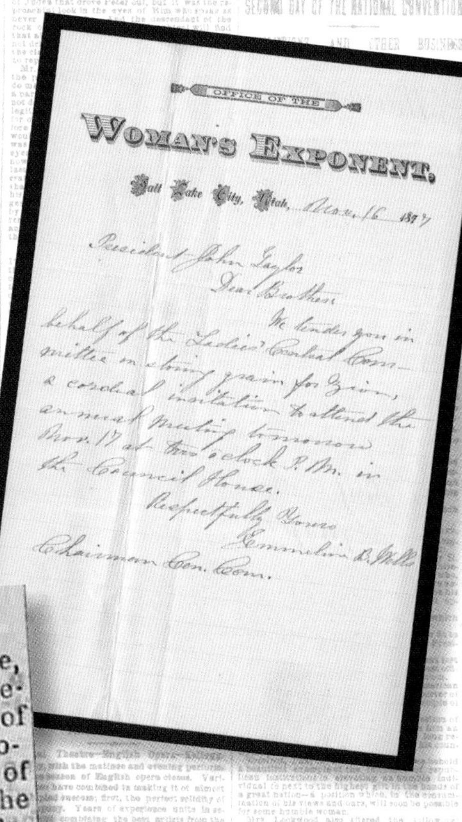

OFFICE OF THE
WOMAN'S EXPONENT,
Salt Lake City, Utah, Nov. 16 1877

President John Taylor
Dear Brother
We tender you in behalf of the Ladies' Central Committee on storing grain for Zion, a cordial invitation to attend the annual meeting tomorrow Nov. 17 at two o'clock P.M. in the Council House.
Respectfully Yours
Emmeline B. Wells
Chairman Cen. Com.

Utah Women Suffrage.

Washington, 28.—In the senate, to-day, Sargent of California, presented a petition of a committee of the National Women's Suffrage association, asking that the women of Utah be protected in exercising the right of the elective franchise. Referred to the committee on territories.

1874 1875 1876

The *Woman's Journal*, an AWSA newspaper edited by Lucy Stone, published a letter from Charlotte Godbe on September 9. Godbe advocated for the extension of women's suffrage and defended women who were the victims of polygamy. Her letter was accompanied by a letter from AWSA leader Caroline Severance endorsing Godbe as the most qualified representative of Utah women.

In response, Emmeline B. Wells countered with her own letter, published in the *Woman's Journal* on October 28, arguing that Charlotte Godbe could not truthfully represent Latter-day Saint women because she had forsaken the faith. This exchange revealed tensions between Wells and Godbe that would continue for years as they vied for leadership of the suffrage movement in Utah.

1877

Emmeline B. Wells became the editor of the *Woman's Exponent* in July. Wells would run the paper for almost forty years, using its pages to communicate Relief Society news, promote women's rights, and establish "a more universal sisterhood" with Eastern suffragists.

"ERE LONG THE WOMEN OF UTAH WILL RISE IN DIGNITY AND STRENGTH, LEAVENING THE WORLD WITH THE IMPORTANCE OF WOMAN SUFFRAGE, TO WHICH THEY HOLD THE KEY."

—*CHARLOTTE GODBE*

THE WOMAN'S EXPONENT.

EMMELINE B. WELLS, - - - - - Editor.

Published semi-monthly, in Salt Lake City, Utah. Terms: One copy one year, $2.00; one copy six months, 1.00. Ten copies for the price of nine. Advertising rates: Each square, ten lines of nonpareil space, one time, $2.00; per month, $3.00. A liberal discount to regular advertisers.

Exponent Office, corner of South Temple, and First East streets, opposite Eagle Gate. Business hours from 10 a.m. to 5 p.m. every day except Sunday.

Address all business communications to
Publishers WOMAN'S EXPONENT,
Salt Lake City, Utah.

SALT LAKE CITY AUGUST 1, 1877.

VALEDICTORY.

In announcing my exit from the editorial department of THE WOMAN'S EXPONENT, I do not feel that I am bidding its patrons and readers "a long and sad farewell;" but hope, as a contributor, to still communicate with them occasionally.

In consideration of the curiosity, and perhaps anxiety, of my friends, personal and otherwise, I give a brief explanation of my reasons for withdrawing from public service for the present. My general health is good, but my head and eyes need recruiting, and I have decided to humor them. I have also decided that during the years of my life which may be properly devoted to the rearing of a family, I will give my special attention to that most important branch of "Home Industry." Not that my interest in the public weal is diminishing, or that I think the best season of a woman's life should be completely absorbed in her domestic duties. But every reflecting mother, and every true philanthropist, can see the happy medium between being selfishly home bound, and foolishly public spirited. At least, I see the interjacence of which I speak, and believe there is wisdom in choosing it. I am content to give up all claim to the EXPONENT, and to be released from further responsibility concerning it, knowing as I do, that I leave it in good hands.

In conclusion, I will say to my sisters old and young, let us subscribe for and write to our little paper. Let us aid all we can to make it more and more interesting and successful in performing its mission.

will and must necessarily precede the setting up of God's kingdom upon the earth.

Women profess to have great intuition, to be particularly susceptible of impressions as regards the future; if such is the case (and we believe it is) then they should be wise in time, and be practical in the movements to become self-sustaining. The women of Utah have been counseled to store up grain for Zion, a mission of which they may justly be proud; a mission such as women never before we think received at any time upon the earth. And now that the harvest is near, and the fields ripening, it is time to take into wise consideration what can be done to carry out this counsel, and make careful and judicious preparation for the future. Parties of ladies could be formed for the especial purpose of gleaning after the reapers. These might be made particularly pleasant, as well as profitable, if the sisters entered into it with that zeal and enthusiasm they do into excursions merely for recreation. Of course the sisters would understand that leave must be obtained from owners of fields, so that there would be no intrusion.

There are very many branches, here in Zion, of home work and industries, in which women might be much more extensively engaged and interested, even without neglecting domestic duties. Women should be quick to comprehend the great advantages to be derived from these home resources, all of which are a part of the Political Economy of Our Mountain Home.

HOME AFFAIRS.

ELDER ORSON PRATT left Salt Lake City for England July 18, and sailed from New York for Liverpool July 31. He has gone thither to superintend the publishing of the Book of Mormon and Doctrine and Covenants in the Phonetic Alphabet, which is about to be introduced into general use in Utah.

AT THIS warm season of the year, bathing is a most healthful and delightful recreation. Salt Lake is preferable to any other place of resort and there is every facility now to make it pleasant and practicable. The Utah Western Railroad have so arranged the time as to suit all who may be anxious to avail themselves of the opportunity of these recuperating and exhilirating baths.

WE WISH to call the attention of the sisters

1877 | 1878 | 1879

1877

In July, Wells sent a letter to the *National Citizen and Ballot Box*, a national suffrage newspaper, volunteering Utah women's help to petition for a constitutional amendment for women's suffrage. Wells declared the "ardent" desire of Utah women "to be one with the women of America in this grand movement."

Relief Society women gathered nearly 7,000 signatures throughout December for NWSA's petition for a constitutional women's suffrage amendment, more than any other state or territory. The proposed amendment would have become the sixteenth amendment to the U.S. Constitution.

1878

On January 28, NWSA leaders Belva Lockwood and Sara Andrews Spencer of Washington, D.C., testified to a congressional committee against proposed legislation to disenfranchise Utah women. Spencer became a bold advocate for Latter-day Saint women's voting rights and developed a close relationship with Emmeline B. Wells.

"THE UNDERSIGNED CITIZENS OF THE UNITED STATES, RESIDENTS OF THE TERRITORY OF UTAH, EARNESTLY PRAY YOUR HONORABLE BODY TO ADOPT MEASURES FOR SO AMENDING THE CONSTITUTION AS TO PROHIBIT THE SEVERAL STATES AND TERRITORIES FROM DISFRANCHISING UNITED STATES CITIZENS ON ACCOUNT OF SEX."

—PETITION FOR WOMAN SUFFRAGE

APPEAL
FOR A
SIXTEENTH AMENDMENT.

Female Suffrage.

Ladies were circulating a petition for signatures yesterday, the object of which was to perpetuate the privileges of women as suffragists in this territory. They are coöperating, in this movement, with the National Suffrage association and have inaugurated the movement with a bright prospect of giving it substantial aid. For about eight years suffrage has been one of Utah's women's rights, and the measures introduced into congress recently, among which is one to deprive them of that right and virtually disfranchise them, has aroused them and induced this protective action. A meeting of ladies is to be held this afternoon, in the 14th ward assembly rooms, at which further steps will be taken in connection with the female suffrage question.

1874 1875 1876

"NEARLY EIGHT YEARS, WE HAVE EXERCISED THE BALLOT WITH OUR OWN FREE WILL AND CHOICE, HAVING FULLY DEMONSTRATED THAT HONORABLE WOMEN COMMAND AS MUCH RESPECT AT THE POLLS, AS IN THE DRAWING-ROOM, THE PARLOR, AND THE CHURCH; AND ALSO, THAT THE PRESENCE OF WOMAN THERE, AS ELSEWHERE, HAS A MORAL AND ELEVATING INFLUENCE."

—MEMORIAL OF THE WOMEN OF UTAH

1877

On March 4, Utah territorial delegate George Q. Cannon introduced a petition to Congress with a few thousand signatures from Latter-day Saint women "plead[ing] in self-defense" to protest an anti-polygamy bill under consideration that would revoke their voting rights.

1878

In the spring of 1878, writer, reformer, and suffragist Lillie Devereaux Blake wrote against "the cruel and absurd" proposal to revoke Utah women's voting rights in order to end polygamy. Because women had no say in electing their congressional representatives, Blake argued that it was unjust for male legislators to make laws restricting women's rights.

Because of her efforts in organizing Utah's petition drive for a constitutional women's suffrage amendment, Emmeline B. Wells was appointed to the NWSA Advisory Committee in May 1878. Ellen Gifford Haydon replaced Annie Godbe as NWSA vice president for Utah.

"THE HANDS OF THE WOMEN OF THE LAND ARE TIED; THEIR VOICES ARE NOT HEARD WITHIN THOSE WALLS; THEY ARE POWERLESS TO PREVENT THIS OR ANY OTHER INSULT. AND YET MEN PRETEND TO REPRESENT US! WHAT AN ABSURDITY! . . . AND NOW, IN ORDER TO ABOLISH POLYGAMY, CONGRESS IS ASKED TO PASS A LAW TO DISFRANCHISE—THE WOMEN!"

—LILLIE DEVEREAUX BLAKE

1879

1878

Emmeline B. Wells sought to amend Utah territorial law to allow women to hold elected office. She was nominated for county treasurer in the Salt Lake County Republican convention by a unanimous vote, but the territorial legislature would not pass a law to remove the word "male" from the qualifications for public office.

In an article on August 15, the *Woman's Exponent* urged foreign-born women in Utah to apply for U.S. citizenship in order to gain voting rights.

NWSA national secretary Sara Andrews Spencer published a letter on October 18 inviting Utah women to the upcoming national suffrage convention in Washington, D.C. She wrote: "Let us by all means have one or more of the enterprising, public-spirited women of Utah present."

Led by Sarah Cooke, Jennie Froiseth, and Cornelia Paddock, about two hundred women who were not Latter-day Saints met in Salt Lake City's Independence Hall on November 7 to form the Ladies' Anti-Polygamy Society of Utah. The Society wrote to First Lady Lucy B. Hayes asking her to support stronger federal anti-polygamy legislation and distributed 30,000 copies of the letter to American clergymen. Women and congregations across the country flooded Congress with petitions asking for stricter anti-polygamy laws.

"IN UTAH, WHERE WOMEN HAVE THE 'FRANCHISE,' IT IS NECESSARY THAT WOMEN SHOULD BE NATURALIZED AS WELL AS MEN. . . . WE WOULD URGE UPON THE SISTERS WHO HAVE NOT ACTED IN THIS MATTER TO DELAY NO LONGER."

—WOMAN'S EXPONENT, *AUGUST 15, 1878*

"WE NEVER THOUGHT THAT WOMAN COULD RISE UP AGAINST WOMAN. . . . IT WILL BE DIAMOND CUT DIAMOND."

—EMMELINE B. WELLS

Latter-day Saint women gathered in the Salt Lake Theatre on November 16 "to protest against the interference of the Anti Polygamy crusade with their rights and privileges as American citizens." The women present resolved to send their "sincere and heartfelt thanks" to NWSA leaders who had defended Utah women's rights. They also resolved to petition Congress with their grievances and printed 10,000 copies of the petition and resolutions.

During the next few months, thousands of Latter-day Saint women held meetings across Utah Territory to affirm the resolutions of the Salt Lake mass meeting and add their signatures to the petition to Congress.

"WE HAVE ASSEMBLED AS WOMEN OF THIS COUNTY, TO PROTEST AGAINST THE INFRINGEMENT OF OUR RELIGIOUS LIBERTIES, AND PROTEST AGAINST ANY MOVEMENT ON THE PART OF CONGRESS TO INJURE OUR FRANCHISE OR THAT OF OUR HUSBANDS."

—MARGARET A. CLUFF

Jennie Anderson Froiseth (1843–1930) *was one of the most prominent and influential opponents of Latter-day Saints in her time. Inspired by her friend, poet, and AWSA leader Julia Ward Howe, Froiseth organized the Blue Tea literary club that established connections among women outside of the Church of Jesus Christ in territorial Utah. She fervently believed in equal suffrage but was a passionate anti-polygamist and felt that Latter-day Saint women should not have voting rights until polygamy was abolished. Froiseth helped establish the Ladies' Anti-Polygamy Society of Utah in 1878, published the* Anti-Polygamy Standard *newspaper, and edited* Women of Mormonism, *which chronicled the trials and suffering of polygamous wives. She was a delegate to the NWSA convention in 1884 and served as NWSA vice president for Utah for several years. Her anti-polygamy feelings were so strong that she refused to participate in organizing the Utah chapter of NWSA created in 1889, instead switching her allegiance to the rival AWSA organization in protest. After Utah women regained suffrage, Froiseth worked with her former rival editor, Emmeline B. Wells, to organize women in the Republican Party of Utah. She also helped establish and preside over the Sarah Daft Home, which continues today as a home for the elderly.*

1880 | 1881 | 1882

Zina D. H. Young (1821–1901) and Zina Young Williams Card (1850–1931) both actively contributed to the suffrage movement in Utah and spoke at national, state, and local suffrage meetings. Zina D. H. Young helped draft the letter to Acting Territorial Governor Mann in 1870 to thank him for signing the suffrage bill, attended a NWSA convention and New York suffrage convention in the winter of 1881–82, and helped organize the Utah Woman Suffrage Association in 1889. She was one of the plural wives of both Joseph Smith and Brigham Young and eventually served as the third President of the Relief Society. Her daughter, Zina Young Williams Card, was one of Brigham Young's daughters and a prominent leader among Utah women in her own right. She spoke at the NWSA convention in 1879 and lobbied Congress and the U.S. president on behalf of Latter-day Saint women's suffrage and religious rights.

1878

New York suffragist Matilda Joslyn Gage, editor of the NWSA's *National Citizen and Ballot Box*, published an article defending Utah women's voting rights called "The Utah Question" in the December 1878 issue.

1879

In *Reynolds v. United States*, the U.S. Supreme Court ruled on January 6 that the First Amendment right to the free exercise of religion did not apply to the practice of polygamy. The ruling upheld the constitutionality of the Morrill Anti-Bigamy Act and paved the way for convictions of polygamists and further anti-polygamy legislation that disenfranchised Utah women.

Emmeline B. Wells and Zina Young Williams attended the annual NWSA convention in Washington, D.C., where they were honored as voting women by riding to the opening session and sitting on the platform with Elizabeth Cady Stanton and Susan B. Anthony. Wells and Williams were appointed to committees and addressed the convention on January 9 and 10, appealing to suffragists across the nation to help Utah women retain their voting rights.

1877 1878 1879

With NWSA leaders Matilda Joslyn Gage and Sara Spencer, Wells and Williams met with President Rutherford B. Hayes on January 13 to urge him to support women's political, social, and civil rights. When Wells and Williams asked President Hayes to protect Latter-day Saint women from anti-polygamy legislation and disenfranchisement, Gage and Spencer quickly clarified that NWSA supported Utah women's voting rights but not polygamy.

On January 17, Wells and Williams testified in favor of women's suffrage before the House of Representatives Judiciary Committee. They spent the next two weeks meeting with legislators and preparing petitions from Latter-day Saint women, which were introduced to the House of Representatives on January 23.

AWSA's *Woman's Journal* published a letter from Amanda Dickinson on March 29 criticizing NWSA for "any appearance of affiliation" with women defending polygamy. AWSA leaders were concerned that such an association could damage public opinion on women's suffrage.

Utah suffragists Mary Godbe, Emmeline B. Wells, and Sarah Ann Cooke were appointed as NWSA officers in May, indicating the organization's attempt to include representatives from the different factions of women in Utah.

1880 1881 1882

"IF GEORGE Q. CANNON CAN SIT IN THE CONGRESS OF THE UNITED STATES WITHOUT COMPROMISING THE BODY . . . I SHOULD THINK MORMON WOMEN MIGHT SIT ON OUR PLATFORM WITHOUT MAKING US RESPONSIBLE FOR THEIR RELIGIOUS FAITH."

—ELIZABETH CADY STANTON

WOMAN'S EXPONENT.

The Rights of the Women of Zion, and the Rights of the Women of all Nations.

SALT LAKE CITY, UTAH, NOVEMBER 1, 1879. Vol. 8. No. 11.

1879

Elizabeth Cady Stanton defended NWSA's alliance with Latter-day Saint suffragists in May, writing: "When the women of a whole territory are threatened with disfranchisement where should they go . . . but to the platform of the National Suffrage Association?"

The *Woman's Exponent* added a motto to its masthead on November 1, 1879: "The Rights of the Women of Zion; and the Rights of the Women of All Nations."

1880

On January 13, Eliza R. Snow, Sarah M. Kimball, Emmeline B. Wells, and fourteen other women petitioned the Utah territorial legislature to allow women to hold public office. The legislature passed a bill doing so on February 18, but federally-appointed Territorial Governor George W. Emery refused to sign it into law.

The Ladies' Anti-Polygamy Society of Utah began publishing the *Anti-Polygamy Standard* in June, with Jennie Froiseth as editor. The newspaper argued that women's suffrage in Utah was "an entirely different matter" than suffrage elsewhere, claiming that Utah women's suffrage perpetuated the practice of polygamy.

1877 1878 1879

The Liberal Party filed a legal challenge to Utah's suffrage law on September 25, seeking to compel the Salt Lake County voting registrar to remove all women's names from the voter list. A "large number of ladies" assembled in the federal courtroom for the hearings. The court upheld women's right to register to vote under the suffrage law.

1881

Zina D. H. Young and Dr. Ellen B. Ferguson traveled east in August for a lecture tour. Church leader Joseph F. Smith was "very favorably disposed about the sisters going East." Ferguson had been an active suffragist in the East before moving to Utah and joining the Church of Jesus Christ.

In October, Zina D. H. Young spoke in Vermont while Dr. Ellen Ferguson lectured in Hartford, Connecticut, with NWSA leader Isabella Beecher Hooker. Hooker's sister Harriet Beecher Stowe, author of *Uncle Tom's Cabin*, invited Ferguson to speak at an upcoming women's convention in Buffalo, New York. However, when Ferguson and Young arrived in Buffalo, they were not permitted to speak because of their religion.

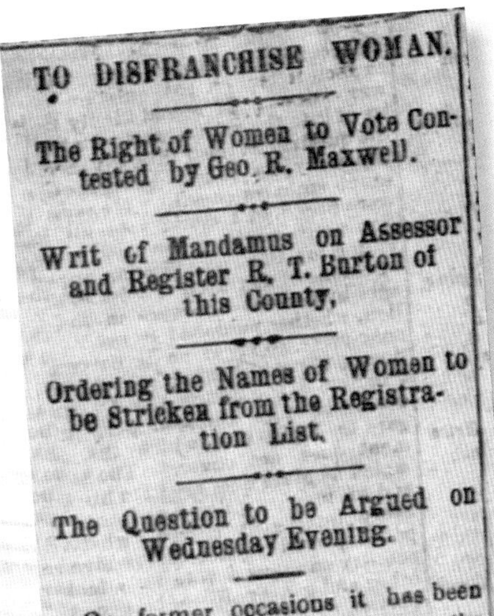

TO DISFRANCHISE WOMAN.

The Right of Women to Vote Contested by Geo. R. Maxwell.

Writ of Mandamus on Assessor and Register R. T. Burton of this County,

Ordering the Names of Women to be Stricken from the Registration List.

The Question to be Argued on Wednesday Evening.

On former occasions it has been stated in these columns that the Liberals had threatened to contest the right of women in this territory to vote. They have publicly averred their determination to make the right of women to vote in this territory one of the points on which they would contest the election of the delegate to Congress—that is, the delegate to be elected in November next, and have assumed very naturally that the Liberal candidate would not be the one. This question of the legality of the law of this territory giving women the right to vote is not a new one, but has never been finally decided by the courts. It was one of the points raised in the

1880 1881 1882

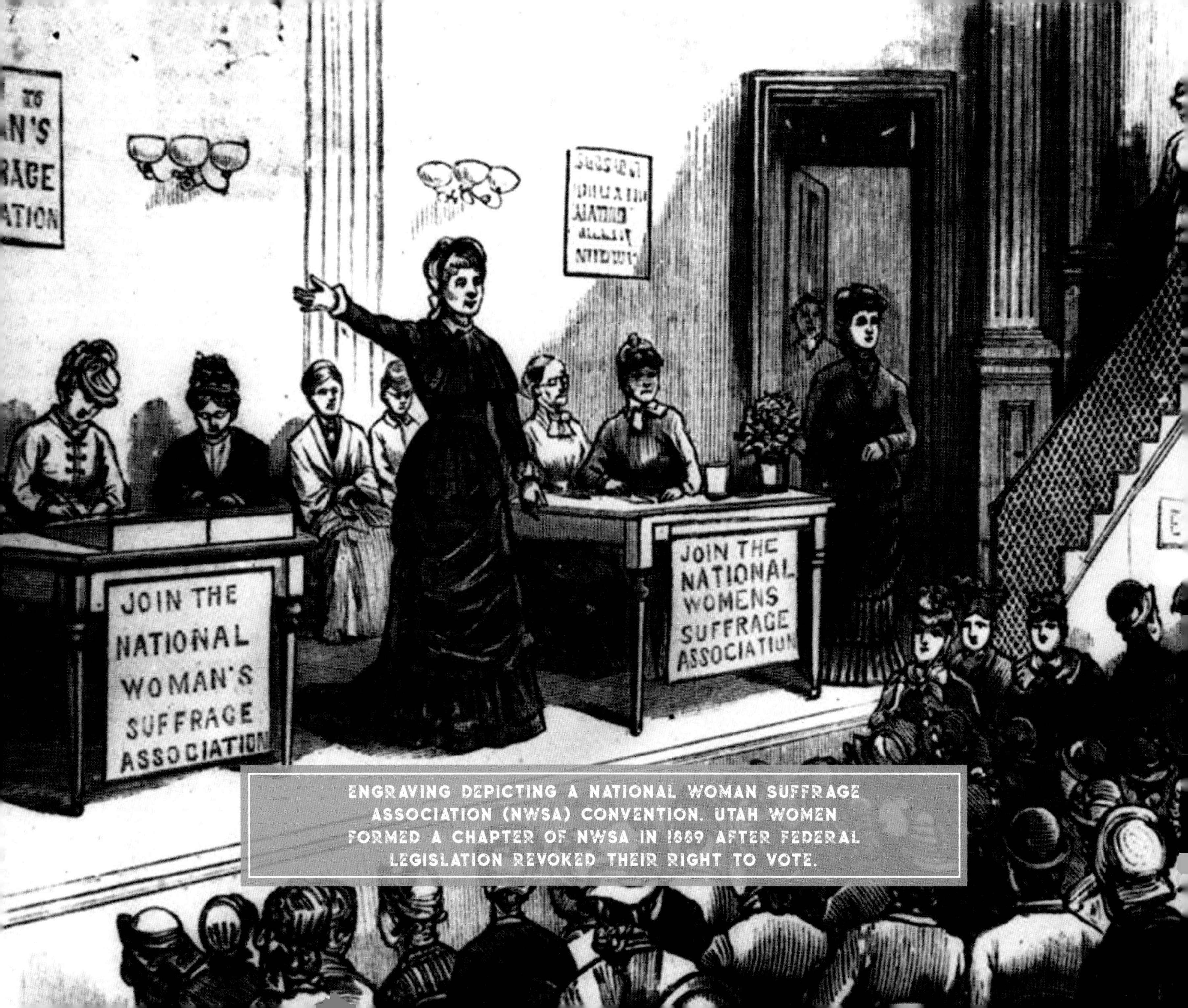

ENGRAVING DEPICTING A NATIONAL WOMAN SUFFRAGE ASSOCIATION (NWSA) CONVENTION. UTAH WOMEN FORMED A CHAPTER OF NWSA IN 1889 AFTER FEDERAL LEGISLATION REVOKED THEIR RIGHT TO VOTE.

LOSING THE VOTE 1882–1889

THE 1880S WERE A TIME OF GROWING OPPOSITION TO both polygamy and Utah women's suffrage rights. Despite petitions and lobbying from Latter-day Saint women, the U.S. Congress passed the Edmunds Act in 1882, which criminalized polygamy and revoked the political and voting rights of polygamous men and women. When this legislation did not sufficiently weaken the political majority in Utah, Congress passed the stronger Edmunds-Tucker Act in 1887, which changed Utah marriage and inheritance laws, replaced local judges to encourage criminal convictions of polygamists, and confiscated assets of the Church. The law also took away the voting rights of all Utah women, whether they were Latter-day Saints or "Gentiles," polygamous or monogamous, married or single.

Throughout this period, most Utah suffragists focused their efforts on defending against the infringement of their religious and civil rights by federal legislation. When their petitions ultimately failed and their rights were revoked in 1887, these suffragists were determined to regain the voting rights they had enjoyed for seventeen years. They mobilized their efforts and worked together to preserve public support for suffrage and to maintain their relationships with national suffrage leaders.

Recognizing the benefits of becoming an official chapter of NWSA, Emily S. Richards and other leading Utah suffragists organized the Utah Woman Suffrage Association (UWSA) in 1889. They used Relief Society networks to establish local suffrage associations in nineteen counties and many more towns throughout the territory. The Utah women who joined these organizations met often to give pro-suffrage speeches, teach lessons in civics, and plan efforts to secure widespread support for including women's suffrage in the state constitution once Utah joined the Union. They also sent delegates to territory-wide and national suffrage conventions.

The anti-polygamy movement of the 1880s had sharply divided suffragists both nationally and within Utah. While Latter-day Saint women almost unanimously defended their right to vote, most other women, including suffragists such as Jennie Froiseth, Cornelia Paddock, and Annie Godbe, refused to become members of the UWSA. Notable exceptions of non-Latter-day-Saint women who joined the UWSA included Lillie Pardee, Margaret Blaine Salisbury, Emma McVicker, Corinne Allen, and Isabelle Cameron Brown. These suffragists made important contributions to the effort to regain the vote in Utah.

1882

Dr. Romania Pratt, Dr. Ellen Ferguson, and Zina D. H. Young attended the New York State Suffrage Association convention in February. Despite being the only enfranchised women there, they encountered escalating anti-polygamy sentiment. Still, one woman, upon examining Zina D. H. Young closely "from top to toe," exclaimed with surprise, "Why, you do not look very degraded!"

Leaders of the Ladies' Anti-Polygamy Society unanimously adopted an "Open Letter to Suffragists" on February 7. While emphasizing their support for the principle of universal suffrage in general, they urged the disenfranchisement of all Latter-day Saint men and women.

Congress passed the Edmunds Act on March 22. The law criminalized polygamy, established a federally-appointed commission to oversee elections in Utah Territory, and disenfranchised polygamous men and women. This removed approximately 12,000 names from the voter rolls, but it still did not erase the political dominance of the People's Party in Utah.

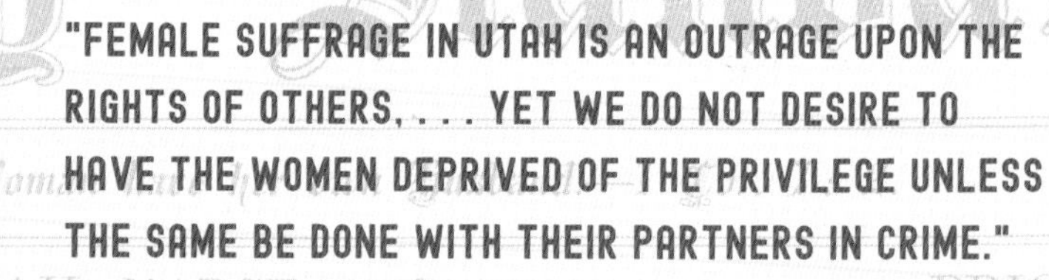

"FEMALE SUFFRAGE IN UTAH IS AN OUTRAGE UPON THE RIGHTS OF OTHERS, . . . YET WE DO NOT DESIRE TO HAVE THE WOMEN DEPRIVED OF THE PRIVILEGE UNLESS THE SAME BE DONE WITH THEIR PARTNERS IN CRIME."
—ANTI-POLYGAMY STANDARD, *MARCH 1882*

"WERE YOU ALWAYS SO EAGER TO HEAR OUR SIDE ON ALL MATTERS OF DISPUTE CONCERNING US AS YOU ARE THE ANTI-SIDE, WE WOULD SUFFER A GREAT DEAL LESS FROM MISREPRESENTATION THAN WE NOW DO."
—*DR. ROMANIA B. PRATT*

Open Letter to the Suffragists of the United States.

(Adopted unanimously at the regular meeting of the Woman's National Anti-Polygamy Society, Feb. 7, 1882.)

DEAR FRIENDS: We women who are active workers in the Anti-polygamy movement in Utah have frequently been censured by some of the Suffrage party for depreciating the fact that the ballot was ever given to the women of this Territory. On this account, we have not always received that fullness of sympathy and co-operation from the party that we had a right to expect from laborers in the same cause—the enfranchisement and elevation of woman. But we believe that you have entirely misunderstood the question, and consequently we desire to explain as fully as possible our position, and the true nature and effects of woman suffrage in Utah.

1882 1883 1884

In appreciation for the Edmunds Act, the women of the Ogden Methodist church made a silk quilt as a present for Vermont Senator Edmunds's wife, Susan. It contained embroidered names of the bill's supporters within Utah and throughout the nation.

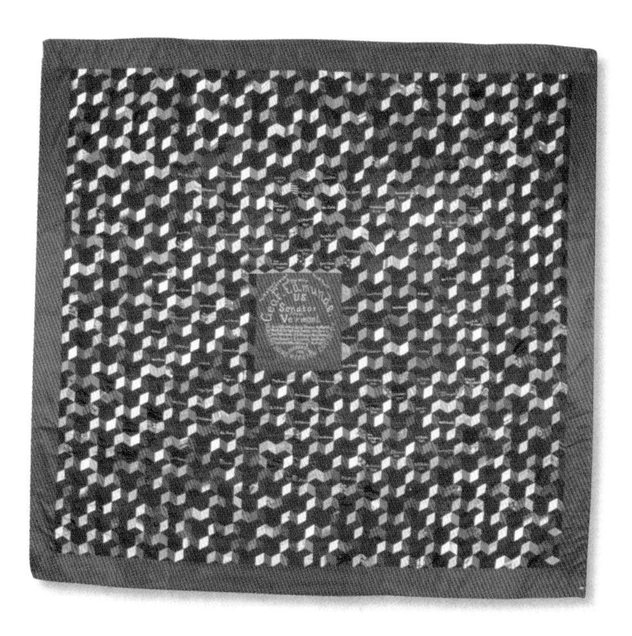

The Utah Commission used registration oaths to decrease Latter-day Saint women's voting in Utah, requiring voters to swear that they had never engaged in a polygamous relationship. Testifying about their marital relationship was offensive and repugnant to Victorian-era women.

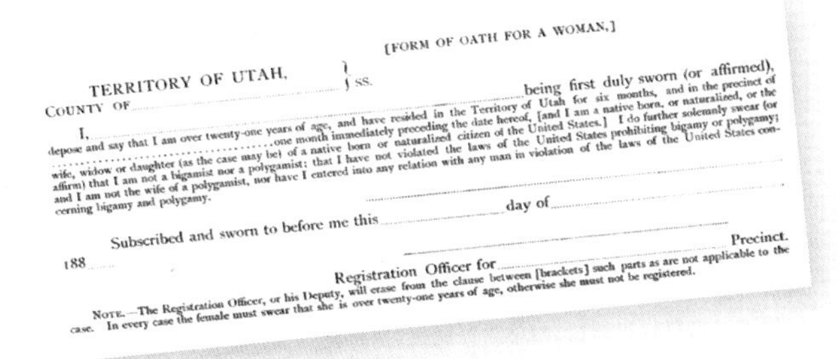

[FORM OF OATH FOR A WOMAN.]

TERRITORY OF UTAH, } ss.
COUNTY OF ....

I, ... being first duly sworn (or affirmed), depose and say that I am over twenty-one years of age, and have resided in the Territory of Utah for six months, and in the precinct of .... one month immediately preceding the date hereof, [and I am a native born, or naturalized, or the wife, widow or daughter (as the case may be) of a native born or naturalized citizen of the United States.] I do further solemnly swear (or affirm) that I am not a bigamist nor a polygamist; that I have not violated the laws of the United States prohibiting bigamy or polygamy; and I am not the wife of a polygamist, nor have I entered into any relation with any man in violation of the laws of the United States concerning bigamy and polygamy.

....

Subscribed and sworn to before me this .... day of .... 188....

....
Registration Officer for .... Precinct.

NOTE.—The Registration Officer, or his Deputy, will erase from the clause between [brackets] such parts as are not applicable to the case. In every case the female must swear that she is over twenty-one years of age, otherwise she must not be registered.

Franklin S. Richards, general counsel for the Church of Jesus Christ, published a legal critique of the Edmunds Act. He argued that it was unconstitutional because it removed civil rights, such as voting, without judicial trial and was applied retroactively, even to plural marriages that had ended before the law was passed.

Franklin S. Richards (1849–1934), a stalwart defender of equal rights, was a crucial advocate on behalf of the women of Utah. He served as county recorder, clerk, and council member in the Utah Territorial Legislature for the district encompassing Weber and Box Elder counties and worked with his wife, UWSA leader Emily S. Richards, to secure women's suffrage when Utah became a state. As a delegate to the 1895 Constitutional Convention, Richards argued for equal suffrage to be included in the new state constitution. His articulate arguments that the same rights must be extended to all people were a decisive factor in the victory for suffrage. During the convention, Richards stated, "Equal suffrage will prove the brightest and purest ray of Utah's glorious star."

1885 1886 1887

1882

Utah held a constitutional convention beginning April 10 to prepare for a sixth attempt to secure statehood. Sarah Kimball, Elizabeth Howard, and Emmeline B. Wells were elected to the committee to frame the proposed state constitution. Elizabeth Farnsworth and Sarah Caroline Shepherd had served as precinct delegates to select Beaver County's representatives at the convention.

The Deseret Hospital was dedicated on July 17 in Salt Lake City. The woman-owned hospital operated for twelve years and was run by a board of prominent Relief Society suffragists and female doctors such as Ellen B. Ferguson, Romania B. Pratt, Martha Hughes Cannon, and Ellis Shipp.

Florence Westcott and the Liberal Party filed test cases to challenge the validity of the women's suffrage law. The People's Party defended women's suffrage. On September 16, Chief Justice of the Territorial Supreme Court John A. Hunter ruled that the act conferring women's suffrage in Utah was legal.

"A WOMAN WHO GOES TO THE POLLS AND DEPOSITS A BALLOT, FEELS HER POLITICAL INDEPENDENCE AND THAT SHE IS VIRTUALLY PART AND PARCEL OF THE GREAT BODY POLITIC, NOT THROUGH HER FATHER OR HUSBAND, BUT IN HER OWN VESTED RIGHT."

—EMMELINE B. WELLS

Anti-polygamy leaders Sarah Ann Cooke and Cornelia Paddock, as well as Relief Society leaders Emmeline B. Wells and Zina D. H. Young, attended the September NWSA Executive Committee meeting in Omaha, Nebraska. Wells presented a report on suffrage in Utah, and Paddock, Wells, and Cooke were appointed to serve as vice presidents for Utah.

The People's Party adopted a platform in its territorial convention on October 12 denouncing efforts to "deprive women voters of the right of suffrage" and stating, "The women of the nation should be endowed with full political freedom."

Following the November 7 election, the *Salt Lake Tribune* lamented the political support that women's votes gave the People's Party: "We believe the Saints polled nearly their full vote; the women seeming determined to exhaust their right."

1885 1886 1887

1883

Belva Lockwood challenged the constitutionality of the Edmunds Act at the January 24 NWSA convention in Washington, D.C. The convention resolved to oppose any further legislation that would disenfranchise even Utah women who were not polygamous or Latter-day Saints.

Senator John A. Logan of Illinois introduced an amendment to the Edmunds Act on February 23, proposing that Congress repeal suffrage rights for all women in Utah. The bill did not pass.

Alice Stone Blackwell, editor of AWSA's *Woman's Journal,* visited Utah in July and was hosted by leaders of the Ladies' Anti-Polygamy Society and suffragists who opposed women's suffrage in Utah. These visits contributed to AWSA's increasing reluctance to defend Utah women's suffrage against congressional attacks.

"THIS PRESENT STRIFE IS A PARTY ONE, AND RAISED FOR POLITICAL EFFECT AND POLITICAL ENDS. THE REAL QUESTION AT ISSUE IS WHETHER UTAH AS A STATE WILL BE DEMOCRATIC OR REPUBLICAN, AND THE HONORABLE SENATOR [EDMUNDS] FROM VERMONT IS ANXIOUS TO CUT OFF DEMOCRATIC VOTES. . . . BY CUTTING OFF THE VOTES OF THE WOMEN, AND AT THE SAME TIME INDULGING A NARROW PREJUDICE AGAINST THE . . . WOMAN'S RIGHTS MOVEMENT."

—*BELVA LOCKWOOD*

1882 | 1883 | 1884

1884

Senator George Hoar from Massachusetts proposed an anti-polygamy bill on January 28, including provisions to repeal all women's suffrage in Utah and to force polygamous wives to testify against their husbands. The bill passed the Senate 33 to 15 but failed to pass the House of Representatives.

At the NWSA convention in Washington, D.C., held March 4–7, Belva Lockwood gave a speech criticizing legislative attempts to disenfranchise the women of Utah. Susan B. Anthony interrupted Lockwood to clarify that NWSA would continue to defend suffrage rights in Utah but would not oppose the other aspects of anti-polygamy legislation.

Angie F. Newman of the anti-polygamy Methodist Episcopal Woman's Home Mission Society visited Utah and worked with Cornelia Paddock to draft a petition to disenfranchise Utah women. They gathered signatures representing 250,000 women throughout the nation.

Indicating increasing tensions over Utah's controversial membership, the 1884 list of NWSA officers omitted Emmeline B. Wells and only included non-Latter-day Saint representatives Sarah Ann Cooke, Cornelia Paddock, and Jennie Froiseth for Utah.

THE DAILY GRAPHIC

"THIS ASSOCIATION MOST EARNESTLY PROTESTS AGAINST NATIONAL INTERFERENCE TO ABOLISH THE RIGHT [TO VOTE] WHEREVER IT HAS BEEN SECURED BY THE LEGISLATURE—AS, FOR EXAMPLE, THE EDMUNDS BILL, WHICH PROPOSES TO DISFRANCHISE ALL THE WOMEN OF UTAH."

—*NATIONAL WOMAN SUFFRAGE ASSOCIATION*

1885 1886 1887

"ALL MORMON WOMEN VOTE WHO ARE PRIVILEGED TO REGISTER. EVERY YOUNG GIRL BORN HERE, AS SOON AS SHE IS TWENTY-ONE YEARS OLD GOES AND REGISTERS AND CONSIDERS IT AS MUCH A DUTY AS TO SAY HER PRAYERS."

—EMMELINE B. WELLS

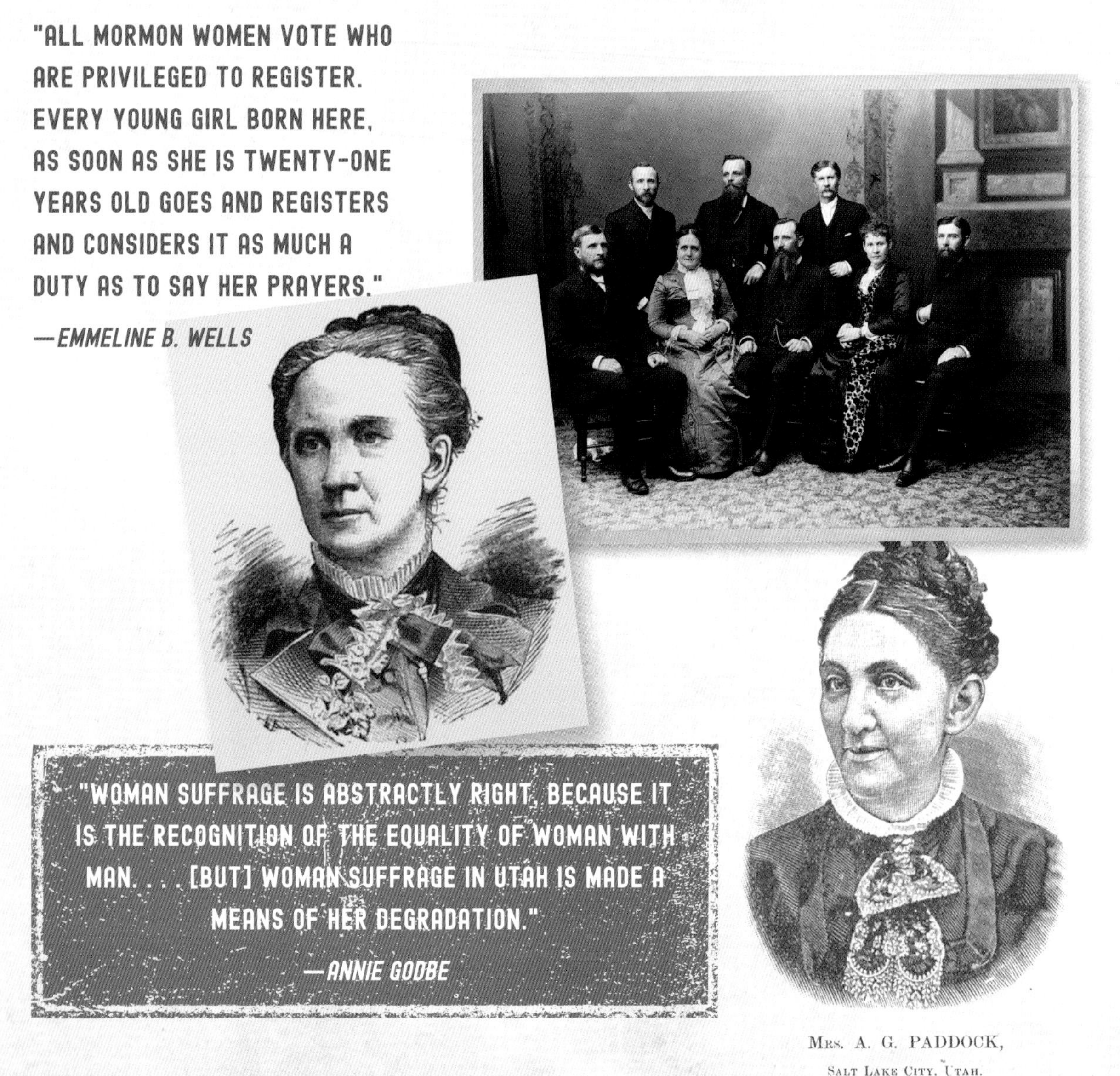

"WOMAN SUFFRAGE IS ABSTRACTLY RIGHT, BECAUSE IT IS THE RECOGNITION OF THE EQUALITY OF WOMAN WITH MAN. . . . [BUT] WOMAN SUFFRAGE IN UTAH IS MADE A MEANS OF HER DEGRADATION."

—ANNIE GODBE

MRS. A. G. PADDOCK,
SALT LAKE CITY, UTAH.

1884

In the sixteenth annual convention report, NWSA published letters from both Emmeline B. Wells and Annie Godbe providing conflicting views on suffrage in Utah. Wells described the benefits of suffrage in Utah and urged NWSA to defend Utah women's right to vote, but Godbe argued that suffrage harmed women by perpetuating polygamy.

1885

Emily S. Richards and Margaret N. Caine, monogamous Latter-day Saint women, attended the NWSA convention in Washington, D.C., in January.

Belva Lockwood visited Utah on a summer lecture tour and received a warm reception from the Latter-day Saint community because of her consistent advocacy on their behalf.

As a result of Cornelia Paddock's petitioning to disenfranchise all Utah women, NWSA removed her name as an officer.

1883 | 1884 | 1885

1886

In Washington, D.C., in February, Emmeline B. Wells met with suffrage leaders and President Grover Cleveland's sister Rose, the acting First Lady, to urge support for Utah women's suffrage rights.

During the NWSA convention in Washington, D.C., on February 17, Indiana suffragist May Wright Sewall reported on the Executive Committee's efforts to fight against the disenfranchisement of Utah women.

Zerelda G. Wallace, an NWSA suffrage leader from Indiana, wrote a personal plea to every member of the U.S. House of Representatives to protect Utah women's right to vote.

The Industrial Christian Home Association of Utah, founded in March by Angie F. Newman, obtained federal funding to provide a refuge for women escaping polygamy. The Industrial Christian Home opened in June 1889, but it closed in 1893 due to a lack of demand and funding.

1886 | 1887 | 1888

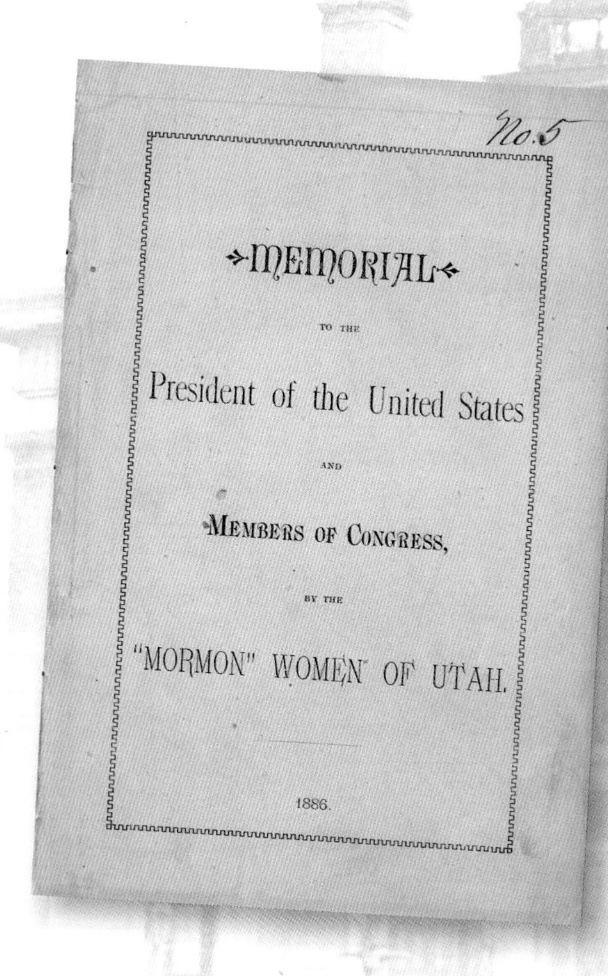
No. 5

MEMORIAL

TO THE

President of the United States

AND

Members of Congress,

BY THE

"MORMON" WOMEN OF UTAH.

1886.

IN THE SENATE OF THE UNITED STATES.

June 8, 1886.—Ordered to lie on the table and be printed.

WOMAN SUFFRAGE IN UTAH.

Mr. Edmunds presented the following petition of Mrs. Angie F. Newman:

To the honorable the Senate and House of Representatives of the Forty-ninth Congress:

Gentlemen: The undersigned respectfully represents that on the 6th of March, 1886, 2,000 Mormon women assembled in mass meeting in Salt Lake City to protest against the invasion of their rights by Federal interference. The spirit of protest crystallized in the appointment of a committee to memorialize the President and Congress of the United States in the form of a "Declaration of grievances" of the *people of Utah*.

Said committee drafted the memorial, proceeded to Washington, and presented it to the national Congress.

It has been read in your hearing.

The following preamble and resolutions are embodied therein:

Whereas, the rights and liberties of women are placed in jeopardy by the present cruel and inhuman proceedings in the Utah courts, and in the *contemplated measures* in Congress to deprive the women voters in Utah of the elective franchise: Therefore, be it

Resolved by the women of Utah in mass meeting assembled, That the suffrage originally conferred upon us as a political privilege has become a vested right by possession and usage for fifteen years, and that we *protest against* being deprived of that right without process of law, and for no other reason than that we do not vote to suit our political opponents.

These resolutions are the expression of Mormon sentiment.

There are Gentile women in Utah Territory. On this basis of civil equality their "protest" should have equal consideration from your honorable body.

As an eye-witness of the ceremonies at the theater meeting; as a long and close observer of the effect of female suffrage in Utah; having definite personal knowledge of the Gentile sentiment touching this question; as a member of the committee who drafted and circulated the petitions (mentioned in the following) asking the repeal of the act conferring the elective franchise upon the women of Utah, the undersigned, in defense of the great principles of civil equity, respectfully submits the following memorial, asking the passage of section 7 of the Edmunds bill, now pending in the national Congress.

All law, human or divine, is the exponent of sovereignty.

The duty of individual or national recognition of or obedience to law is measured by the nature of that sovereignty.

1886

More than 2,000 Latter-day Saint suffragists held a protest in the Salt Lake Theatre on March 6 and drafted a petition against further anti-suffrage legislation.

Dr. Ellen Ferguson and Emmeline B. Wells presented the Latter-day Saint women's petition to Congress and to President Grover Cleveland in Washington, D.C.

Angie F. Newman presented a counter-petition to Congress on June 8 on behalf of the "Gentile" women of Utah, calling for the repeal of women's suffrage there.

"WE PROTEST AGAINST THE MOVEMENT TO DEPRIVE US OF THE ELECTIVE FRANCHISE, WHICH WE HAVE EXERCISED FOR OVER FIFTEEN YEARS. . . . OUR ONLY CRIME IS THAT WE HAVE NOT VOTED AS OUR PERSECUTORS DICTATE."

—MEMORIAL BY THE MORMON WOMEN OF UTAH

1884 1885 1886

1887

At the NWSA convention in Washington, D.C., on January 25, a delegation of twenty women petitioned President Cleveland to veto the proposed disenfranchisement of Utah women and call for a congressional hearing.

Prominent Latter-day Saint suffragists sent a telegram to Susan B. Anthony in January, thanking NWSA leaders for "exercising their influence" to oppose the "obnoxious" Edmunds-Tucker bill.

The Edmunds-Tucker Act became law on March 3, repealing women's suffrage and disenfranchising all Utah women no matter their religion or marital status. The federal legislation also confiscated Church property, disincorporated the Church, changed marriage and inheritance laws, revoked civil liberties for polygamists, and increased imprisonment of polygamists by replacing local judges with federally-appointed judges.

Thanks to the Woman Suffragists. —On Saturday evening the following telegram was sent from this city to Washington:

Miss Susan B. Anthony, Riggs House, Washington D. C.:

In behalf of the women of Utah, we, the undersigned, respectfully tender our grateful acknowledgments to the officers and members of the National Woman's Suffrage Convention, for the able and timely effort made in exercising their influence with the President and the people of the United States in opposing the disfranchisement of the legal women voters of Utah and the obnoxious Edmund's-Tucker bill.

Signed

E. R. Snow Smith, M. Isabella Horne, Jane S. Richards, Josephine R. West, Zina Y. Williams, Zina D. H. Young, Sarah M. Kimball, Emily S. Richards, E. B. Ferguson, M. D., R. B. Pratt, M. D., Emmeline B. Wells.

"IT SHALL NOT BE LAWFUL FOR ANY FEMALE TO VOTE AT ANY ELECTION HEREAFTER HELD IN THE TERRITORY OF UTAH."

—EDMUNDS-TUCKER ACT

1888

NWSA organized the first International Council of Women (ICW) and National Council of Women (NCW) on March 25. This coalition of important women's organizations sought to create "universal sisterhood." Utah delegates to this founding convention included both Latter-day Saint and anti-polygamy women.

Utah suffragists Isabelle Cameron Brown, Jennie Froiseth, and Emily S. Richards attended the April 3 NWSA executive session and received authorization to form a Utah territorial suffrage association. Despite opposition to Utah's inclusion, Brown and Richards were appointed NWSA vice presidents for Utah and a resolution was passed to accept "women of all classes, all races and all religions" as NWSA members.

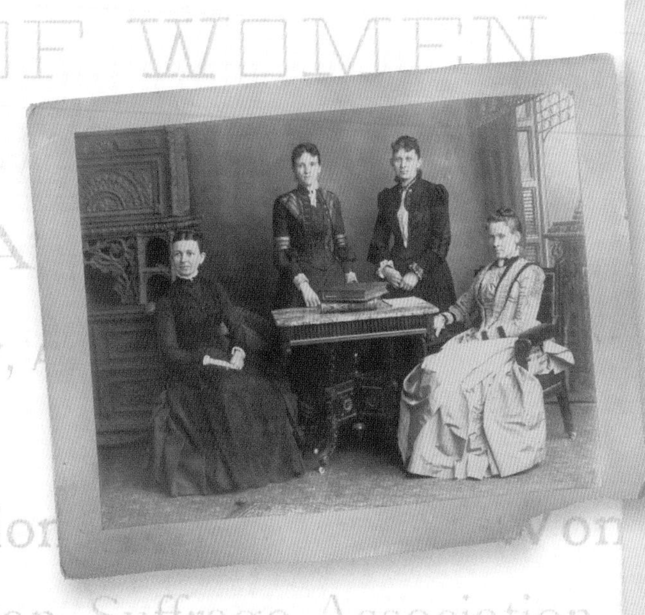

PLEASE READ RAILROAD COMMITTEE NOTICE ON PAGE 2.
SEVENTH EDITION OF 5,000.

INTERNATIONAL
Council of Women
ASSEMBLED BY THE
National Woman Suffrage Association
OF THE UNITED STATES,
TO CELEBRATE THE
FORTIETH ANNIVERSARY
OF THE
FIRST WOMAN'S RIGHTS CONVENTION.

ALBAUGH'S OPERA HOUSE,
WASHINGTON, D. C.,
March 25th to April 1st, 1888, inclusive.

SUNDAY, MARCH 25, 2:30 P. M.—RELIGIOUS SERVICE.

March 26, 10 A. M.—FORMAL OPENING OF COUNCIL.
7:45 P. M.—EDUCATION.
March 27, 10 A. M.—PHILANTHROPIES.
7:45 P. M.—TEMPERANCE.
March 28, 10 A. M.—INDUSTRIES.
7:45 P. M.—PROFESSIONS.
March 29, 10 A. M.—ORGANIZATION.
7:45 P. M.—LEGAL CONDITIONS.
March 30, 10 A. M.—SOCIAL PURITY. (Session for Women Alone.)
7:45 P. M.—POLITICAL CONDITIONS.
March 31, 10 A. M.—PIONEERS' CONFERENCE.
7:45 P. M.—POLITICAL CONDITIONS.

SUNDAY, APRIL 1, 2:30 P. M., RELIGIOUS SYMPOSIUM.

SEASON TICK... ...TS.
SEASON TICK... ...TS.

HISTORY AND MINUTES
OF THE
NATIONAL COUNCIL OF WOMEN
OF THE UNITED STATES,
ORGANIZED IN WASHINGTON, D. C., MARCH 31, 1888.
EDITED BY
LOUISE BARNUM ROBBINS,
Corresponding Secretary.
BY ORDER OF THE EXECUTIVE COMMITTEE OF THE COUNCIL.

BOSTON, MASS.:
E. B. STILLINGS & CO., 55 SUDBURY STREET.
1898.

"WE, WOMEN OF THE UNITED STATES, SINCERELY BELIEVING THAT THE BEST GOOD OF OUR HOMES AND NATION WILL BE ADVANCED BY OUR OWN GREATER UNITY OF THOUGHT, SYMPATHY AND PURPOSE, . . . DO HEREBY BAND OURSELVES TOGETHER IN A CONFEDERATION OF WORKERS COMMITTED TO THE OVERTHROW OF ALL FORMS OF IGNORANCE AND INJUSTICE."

—PREAMBLE TO THE CONSTITUTION OF THE NATIONAL COUNCIL OF WOMEN

AWSA leaders including Julia Ward Howe and Lillie Devereaux Blake visited Salt Lake City that summer. They met with Jennie Froiseth and others but refused to lecture to Latter-day Saint suffragists. After repeated requests, Howe reluctantly agreed to meet privately with Emmeline B. Wells.

NWSA leaders Clara Bewick Colby and Elizabeth Lyle Saxon visited Salt Lake City in September to show support for Utah women's suffrage. They spoke in the Salt Lake Theatre and Assembly Hall, attended a reception with over 500 people at the Gardo house, and visited with Latter-day Saint suffragists as well as anti-polygamy women.

1887 | 1888 | 1889

1888

Jennie Froiseth and Cornelia Paddock, both former NWSA vice presidents for Utah, joined the more conservative AWSA in protest of NWSA's formal inclusion of polygamous suffragists from Utah. At the annual AWSA convention held November 20–22, Froiseth and Paddock were appointed as the first AWSA officers from Utah.

1889

Relief Society leaders Emmeline B. Wells, Sarah M. Kimball, Jane S. Richards, Zina D. H. Young, Emily S. Richards, and Bathsheba W. Smith met with Church leaders on January 2 to discuss organizing a territorial suffrage association. Church president Wilford Woodruff approved and recommended that Emily S. Richards attend the upcoming NWSA conference.

Isabelle Cameron Brown, Emmeline B. Wells, and Emily S. Richards requested Jennie Froiseth's help in organizing a Utah chapter of NWSA where "all classes of women in the Territory would co-operate." Froiseth refused to be involved because of her opposition to Utah women's suffrage while polygamy still existed.

Mrs. Jennie Anderson Froiseth.

1884 1885 1886

"'TIS BETTER TO HAVE VOTED AND BEEN DISFRANCHISED, THAN TO NEVER HAVE VOTED AT ALL."

—EMILY S. RICHARDS

The Utah Woman Suffrage Association (UWSA) officially organized on January 10 in Salt Lake City's Assembly Hall with a leadership of monogamous women. Margaret N. Caine was unanimously elected as president, Lydia D. Alder, Nellie Webber, and Priscilla Jennings Riter as vice presidents, Cornelia H. Clayton as secretary, Margaret Dwyer as treasurer, and Charlotte Godbe Kirby as corresponding secretary.

At the January NWSA convention in Washington, D.C., Emily S. Richards reported that the fledgling UWSA already had 200 dues-paying members. She also presented 8,393 signatures of suffrage supporters in Utah.

Emily S. Richards (1850–1929) was a standout speaker, organizer, and public face of Utah suffragists for many years. She lived for a time in Washington, D.C., where she lobbied on behalf of Utah women. Richards co-organized the Utah Woman Suffrage Association in 1889 and established local branches throughout Utah. In 1895, she campaigned alongside her husband, Franklin S. Richards, to ensure that the Utah state constitution would include equal suffrage. She spoke at national and international suffrage conventions, testified before Congress, was a delegate at the International Council of Women conferences in Berlin and Toronto, and led post-statehood suffrage activism as the President of the Utah Council of Women.

1887 | 1888 | 1889

EVERY LADY SHOULD FEEL IT HER DUTY TO MAKE AN EFFORT TO OBTAIN THE FRANCHISE. MANY DO NOT UNDERSTAND THE TRUE MEANING OF WOMAN SUFFRAGE. SOME THINK WOMAN IS TRYING TO USURP MAN'S RIGHTS. NOT SO! SHE ONLY DESIRES TO STAND SIDE BY SIDE WITH HIM, AND SHARE THOSE PRIVILEGES HE VALUES AS INESTIMABLE."

—ELIZABETH ANN SCHOFIELD

W. S. A. IN JUAB COUNTY.

In answer to a call made by a few of the most inflnential women of Juab County, for a meeting to be held, a large number of citizens both ladies and gentlemen, assembled at the tabernacle, Nephi, at 2 p.m., March 11, 1889, to organize a Suffrage Association. Miss L. A. Schofield nominated Mrs. M. Pitchforth to act as chairman. Seconded by several ladies, after which the motion was put and carried. On motion of Mrs. A. L. Bigler, Mrs M. E. Teasdale was nominated Secretary of the meeting; seconded and carried· Mrs. S. A. Andrews then offered prayer. The Secretary then read the call. Miss L. A. Schofield was then elected President by unanimous vote. Mrs. E. B. Bryan was elected First Vice Presidenst Mrs. M. E. Neff Second Vice President; Mr;. M. A. Cazier third Vice President; Miss Louie Udall, Secretary; Mrs. Annie L. Atkin, Recording, and Corresponding Secretary; Mrs. L. A. Hartley, Treasurer.

The following named ladies were elected by unanimous vote, to act as an Executive Board: Mrs. M. A. Grover, chairman; Mrs. M. E. Whitmore, Mrs. Sarah Abbott, Mrs. W. H. Jones, Miss Emma Adams, Mrs. Mercy Wright, nd Mrs. Julia Paxman. Miss Louie Udall read the Constitution of the National Woman Suffrage Association; after which a motion was made that the same be adopted. Seconded and carried by unanimous vote. Mrs. Annie Atkins read the By-Laws of the Salt Lake

1889

The UWSA held a meeting in the Social Hall on February 9 with speeches by Lydia D. Alder, James E. Talmage, and Camilla Cobb. Charlotte Godbe read Emily S. Richards's speech to the NWSA convention.

The Beaver County Woman Suffrage Association formed on February 16, with Julia P. M. Farnsworth as the first president. She declared her commitment to both men and women: "They are inseparable; . . . neither alone can form a perfect home, community, or nation. . . . God grant that . . . the subject of woman suffrage [will be] properly adjusted, and free and equal rights given to all." The association later began to publish a local monthly suffrage bulletin called the *Equal Rights Banner*.

The Juab County Woman Suffrage Association was organized in the Nephi Tabernacle on March 11. Elizabeth "Lizzie" Schofield was unanimously elected president.

1884 | 1885 | 1886

No 25.

CERTIFICATE OF MEMBERSHIP
OF THE
Woman Suffrage Association
OF
SEVIER COUNTY, UTAH,

This is to Certify that [illegible] having subscribed and paid for membership in the Woman Suffrage Association of Sevier County, is entitled to all its benefits and advantages for one year from the date hereof.

[illegible] Secretary. [illegible] President.

DATE [illegible]

N. B.—This Certificate is not Transferable.

The UWSA held a large meeting in the Salt Lake City Assembly Hall on April 11, with speeches by Bishop Orson F. Whitney, the Hon. Charles W. Penrose, the Hon. George Q. Cannon, Dr. Martha Hughes Cannon, Zina D. H. Young, Emily S. Richards, Ida Snow Gibbs, and Nellie R. Webber.

By May, the UWSA had established fourteen county branches. In addition to the territorial association's conventions and events, the county organizations held monthly meetings and actively advocated for suffrage on a local level.

The Morgan County Woman Suffrage Association was organized on June 22 in a mass meeting at the Morgan City schoolhouse. Hulda Cordelia Smith, the new president, stated that she "was enlisted in the Woman's cause, [and] never could see why women should be subject to laws she had no hand in making, nor ruled over by those not of her choice."

Margaret N. Caine resigned as president of the UWSA in July, and vice president Lydia D. Alder served as interim president for the rest of the year.

1887 1888 1889

UTAH SUFFRAGISTS ELMINA S. TAYLOR AND EMMELINE B. WELLS (SEATED FIRST AND SECOND FROM LEFT) AND SUSAN B. ANTHONY (SEATED SECOND FROM RIGHT) JOINED OTHER OFFICERS AND DELEGATES OF THE NATIONAL COUNCIL OF WOMEN IN FEBRUARY 1895 AT A CONFERENCE IN WASHINGTON, D.C.

4 REGAINING THE VOTE 1890–1895

THE YEAR 1890 MARKED A NEW ERA OF COOPERATION AMONG women in the national suffrage movement and in Utah. Although underlying tensions remained, both the merger of the rival suffrage organizations into the National American Woman Suffrage Association (NAWSA) and the Church's Manifesto that officially ended the practice of polygamy paved the way for a "more universal sisterhood" and for statehood. Latter-day Saint women earnestly sought to establish common ground with women across the nation. As a symbol of this unity, nearly twenty Utah women participated in the "World's Congress of Representative Women" at the Chicago World's Fair in 1893.

Though Utah Territory had unsuccessfully applied for statehood several times over the previous four decades, Congress passed the Enabling Act in 1894, essentially inviting Utah to apply again. With statehood on the horizon, Utah's suffragists joined forces and worked to ensure that the new state constitution would reinstate women's suffrage.

At Utah's 1895 constitutional convention, male delegates arrived intending to include a clause in the state constitution granting Utah women the right to vote and hold public office. Davis County delegate Brigham H. Roberts unexpectedly raised arguments against the suffrage clause, but a large majority of delegates, led by Franklin S. Richards and Orson F. Whitney, supported women's right to vote. UWSA members tirelessly lobbied for the inclusion of suffrage in the constitution and argued against a rising call for separate consideration of the suffrage issue after statehood. Women in Utah submitted petitions on both sides of the question, and the pro-suffrage delegates carried the day. The proposed state constitution included an equal suffrage clause.

Less than a month later, NAWSA leaders Susan B. Anthony and Dr. Anna Howard Shaw presided at the Rocky Mountain suffrage convention in Salt Lake City, speaking to a crowd of 6,000 in the Salt Lake Tabernacle. Delegates from Wyoming, Colorado, Idaho, and Montana gathered to celebrate Utah suffragists' success, learn from their experience, and plan for suffrage campaigns in other states. The conference provided a moment of unity among Utah women who had come together to work for the common goal of suffrage after the bitter divisions over polygamy.

Sarah M. Kimball (1818–1898) was influential in establishing the first Female Relief Society in Nauvoo, Illinois, and later wrote, "The sure foundations of the suffrage cause were deeply and permanently laid on the 17th of March, 1842," the day the Relief Society was first organized. Kimball served as the president of the Salt Lake 15th Ward Relief Society for forty-two years and oversaw the building of the 15th Ward Relief Society Hall, where the first organized effort to secure Utah women's suffrage occurred. A self-proclaimed "women's rights woman," Kimball served as president of the UWSA and as a delegate to annual NAWSA conventions.

1890

At its first annual convention, held on January 11 in the Salt Lake City Social Hall, the Utah Woman Suffrage Association (UWSA) elected Sarah M. Kimball as president, Emily S. Richards as vice president, and Phebe Young Beatie as chair of the executive committee.

NWSA and AWSA merged to form the National American Woman Suffrage Association (NAWSA) on February 18, with Elizabeth Cady Stanton as president. Despite internal disagreements over whether to include Latter-day Saint women, NAWSA accepted the membership of the Utah Woman Suffrage Association.

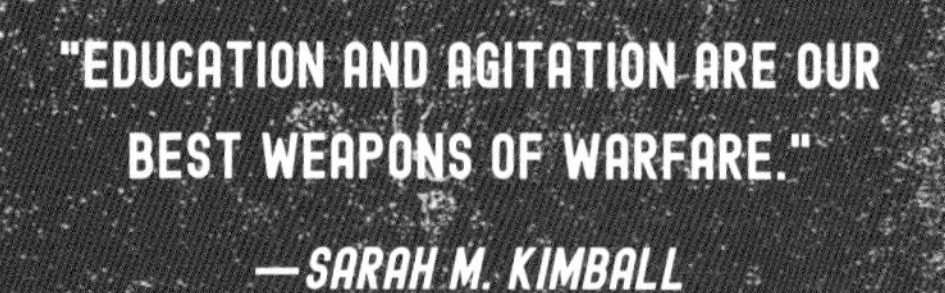

1890 | 1891 | 1892

Sarah M. Kimball and Maria Young Dougall represented Utah at the first convention of the newly-merged NAWSA in Washington, D.C., held February 18–21. They reported that the UWSA had three hundred dues-paying members in addition to sixteen county suffrage organizations.

Wyoming was granted statehood on July 23 and became the first state with women's suffrage. The UWSA assembled in Salt Lake City's Liberty Park to "rejoice in the good fortune of Wyoming women," and sent a telegram of congratulations to Wyoming suffrage leader Amalia Post.

Church President Wilford Woodruff publicly issued a manifesto on September 25 that officially discontinued the practice of plural marriage. The manifesto was formally accepted by the membership of the Church of Jesus Christ at general conference on October 6.

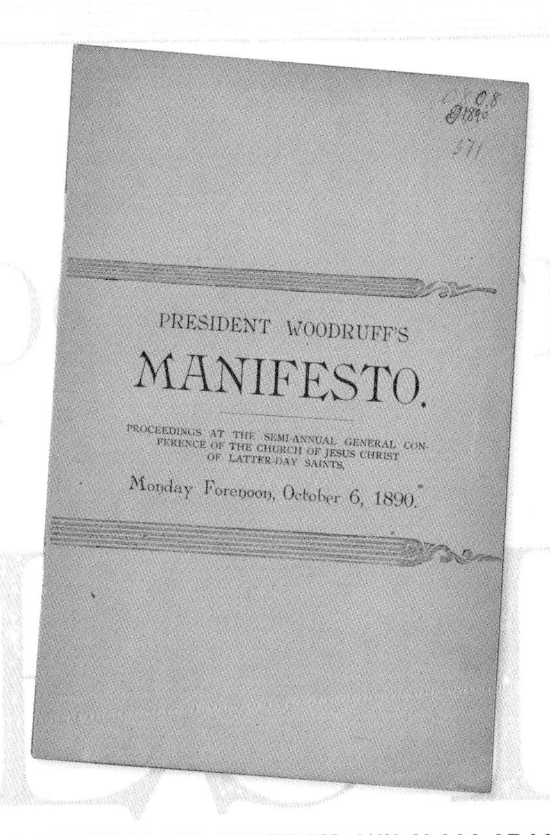

PRESIDENT WOODRUFF'S

MANIFESTO.

PROCEEDINGS AT THE SEMI-ANNUAL GENERAL CONFERENCE OF THE CHURCH OF JESUS CHRIST OF LATTER-DAY SAINTS.

Monday Forenoon, October 6, 1890.

"WHEREVER AND WHATEVER ANY CLASS OF WOMEN SUFFER . . . A VOICE IN THEIR BEHALF SHOULD BE HEARD IN OUR CONVENTIONS. . . . COLORED WOMEN, INDIAN WOMEN, MORMON WOMEN . . . HAVE BEEN HEARD IN THESE WASHINGTON CONVENTIONS & I TRUST THEY ALWAYS WILL BE."

—*ELIZABETH CADY STANTON*

1893 1894 1895

1891

Utah sent ten delegates to the annual NAWSA convention in Washington, D.C., on February 26. The UWSA was second only to New York in its number of suffrage association members.

The Liberal and People's parties were dissolved by June as local Republican and Democrat political parties formed to prepare for statehood and align Utah more fully with the rest of the nation. Partisan politics created both bridges and divisions among Utah women.

Two Latter-day Saint women's organizations, the Relief Society and the Young Ladies' Mutual Improvement Association, were charter members of the National Council of Women at its first convention, held in Washington, D.C., February 23–25. The Relief Society and YLMIA continued active membership in this influential coalition of women's organizations until 1987.

"[EMMELINE B. WELLS'] ADVOCACY OF WRONGED WOMEN AND THE EQUALITY OF THE SEX HAS BEEN PARTICULARLY FEARLESS."

—WASHINGTON POST, *FEBRUARY 23, 1891*

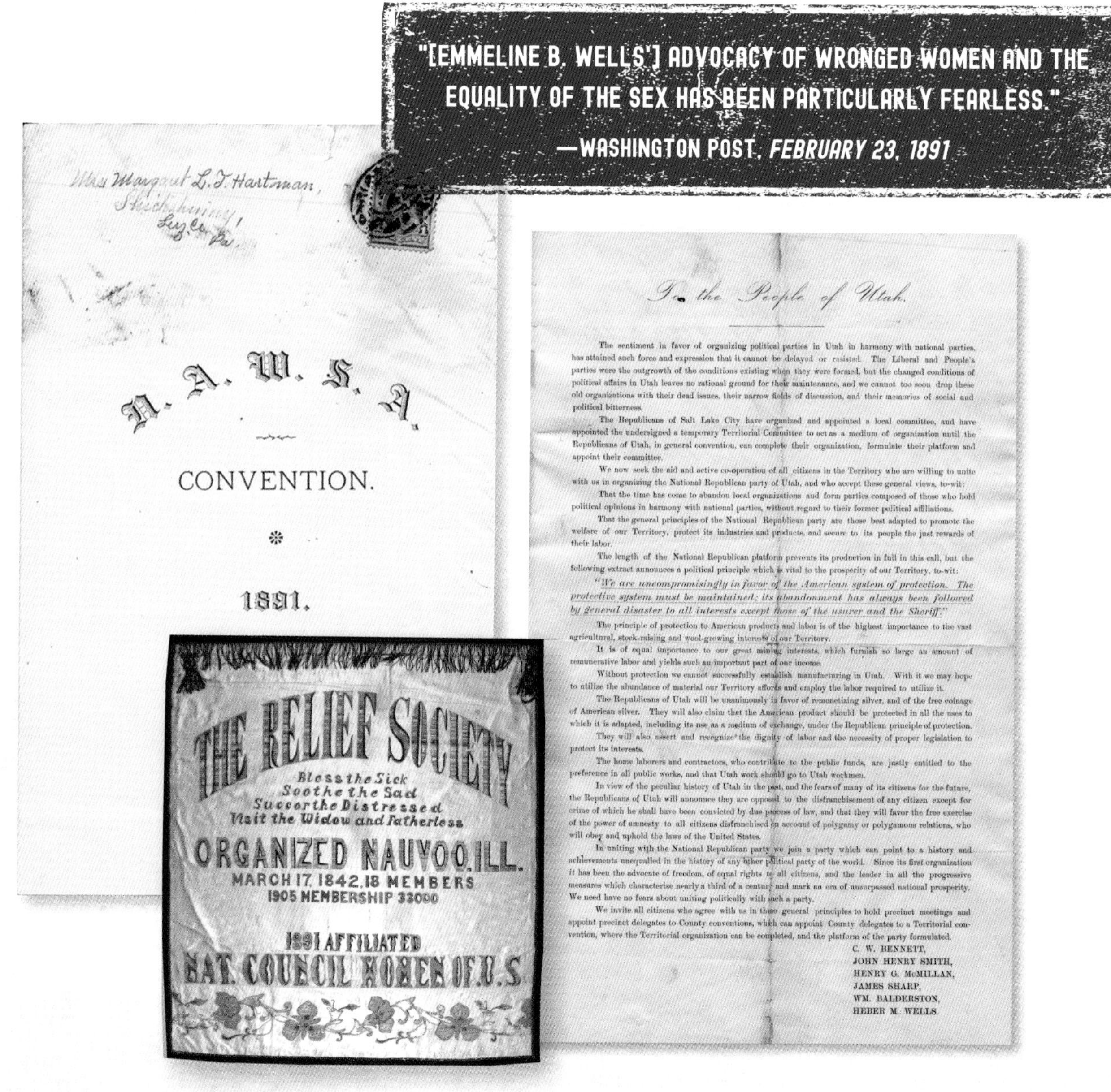

To the People of Utah.

The sentiment in favor of organizing political parties in Utah in harmony with national parties, has attained such force and expression that it cannot be delayed or resisted. The Liberal and People's parties were the outgrowth of the conditions existing when they were formed, but the changed conditions of political affairs in Utah leaves no rational ground for their maintenance, and we cannot too soon drop these old organizations with their dead issues, their narrow fields of discussion, and their memories of social and political bitterness.

The Republicans of Salt Lake City have organized and appointed a local committee, and have appointed the undersigned a temporary Territorial Committee to act as a medium of organization until the Republicans of Utah, in general convention, can complete their organization, formulate their platform and appoint their committee.

We now seek the aid and active co-operation of all citizens in the Territory who are willing to unite with us in organizing the National Republican party of Utah, and who accept these general views, to-wit:

That the time has come to abandon local organizations and form parties composed of those who hold political opinions in harmony with national parties, without regard to their former political affiliations.

That the general principles of the National Republican party are those best adapted to promote the welfare of our Territory, protect its industries and products, and secure to its people the just rewards of their labor.

The length of the National Republican platform prevents its production in full in this call, but the following extract announces a political principle which is vital to the prosperity of our Territory, to-wit:

"*We are uncompromisingly in favor of the American system of protection. The protective system must be maintained; its abandonment has always been followed by general disaster to all interests except those of the usurer and the Sheriff.*"

The principle of protection to American products and labor is of the highest importance to the vast agricultural, stock-raising and wool-growing interests of our Territory.

It is of equal importance to our great mining interests, which furnish so large an amount of remunerative labor and yields such an important part of our income.

Without protection we cannot successfully establish manufacturing in Utah. With it we may hope to utilize the abundance of material our Territory affords and employ the labor required to utilize it.

The Republicans of Utah will be unanimously in favor of remonetizing silver, and of the free coinage of American silver. They will also claim that the American product should be protected in all the uses to which it is adapted, including its use as a medium of exchange, under the Republican principle of protection.

They will also assert and recognize the dignity of labor and the necessity of proper legislation to protect its interests.

The home laborers and contractors, who contribute to the public funds, are justly entitled to the preference in all public works, and that Utah work should go to Utah workmen.

In view of the peculiar history of Utah in the past, and the fears of many of its citizens for the future, the Republicans of Utah will announce they are opposed to the disfranchisement of any citizen except for crime of which he shall have been convicted by due process of law, and that they will favor the free exercise of the power of amnesty to all citizens disfranchised on account of polygamy or polygamous relations, who will obey and uphold the laws of the United States.

In uniting with the National Republican party we join a party which can point to a history and achievements unequalled in the history of any other political party of the world. Since its first organization it has been the advocate of freedom, of equal rights to all citizens, and the leader in all the progressive measures which characterize nearly a third of a century and mark an era of unsurpassed national prosperity. We need have no fears about uniting politically with such a party.

We invite all citizens who agree with us in these general principles to hold precinct meetings and appoint precinct delegates to County conventions, which can appoint County delegates to a Territorial convention, where the Territorial organization can be completed, and the platform of the party formulated.

C. W. BENNETT,
JOHN HENRY SMITH,
HENRY G. McMILLAN,
JAMES SHARP,
WM. BALDERSTON,
HEBER M. WELLS.

1890 | 1891 | 1892

"CLAIM THE BIRTHRIGHT OF THE FREE—EQUAL RIGHTS AND LIBERTY."

—*EMILY WOODMANSEE*

"GREETINGS, DEAR FRIENDS: THAT OUR CITIZENS' RIGHT TO VOTE MAY SOON BE SECURED IS THE PRAYER OF YOUR CO-WORKER."

—*SUSAN B. ANTHONY*

The UWSA published the Suffrage Songbook, a collection of nineteen songs about suffrage and liberty. The songs were sung at suffrage meetings and events, often to the melodies of Latter-day Saint hymns.

1892

The UWSA and several members of the legislature celebrated Susan B. Anthony's birthday on February 15 in one of the largest halls in Salt Lake City. Anthony responded with a telegram.

A large suffrage rally in American Fork, Utah, on July 29 featured yellow as the color of suffrage: "Ladies wore the yellow ribbon and many gentlemen the sunflower . . . and the entire day was a grand feast of suffrage sentiment." Local associations often held social events to raise money and awareness for the cause.

1893

U.S. President Benjamin Harrison granted amnesty to Latter-day Saints on January 4 for the penalties imposed by anti-polygamy legislation. The presidential pardon restored voting rights to polygamous men, but it did not give suffrage back to Utah women.

1893 | 1894 | 1895

The World's Congress of Representative Women, *part of the six-month World's Fair in Chicago, provided a significant platform for Utah women to present themselves to the world as refined, educated, and articulate advocates for women. They joined more than 600 women representing over 100 organizations and 33 countries to give speeches and presentations that drew an estimated 150,000 visitors. As members of the International Council of Women (ICW), Utah's Relief Society and YLMIA were invited to conduct their own sessions, and Emmeline B. Wells presided at the General Congress of the ICW held at the Palace of Fine Arts. The Woman's Building and Utah's own state building featured exhibits of Utah silk and handiwork, statistical surveys of Utah women's work and achievements, and examples of Utah women's literary accomplishments. The extensive preparations for this seminal event drew Utah women together despite religious differences, healed breaches with national suffrage leaders, and solidified support for their work to regain suffrage rights in Utah.*

1893

A large delegation of almost twenty Utah women participated in the World's Congress of Representative Women, which was held May 15–21 as part of the larger Columbian Exposition at the World's Fair in Chicago.

Dr. Martha Hughes Cannon delivered a speech at the World's Congress about Utah women's loyalty and sacrifice in building up the territory. The *Chicago Record* concluded: "Mrs. Dr. Martha Hughes Cannon . . . is considered one of the brightest exponents of woman's cause in the United States."

"WITH ALL THAT WEALTH AND CIVILIZATION CAN GIVE FOR THE ADVANCEMENT OF SCIENCE, LITERATURE, AND ART, WHAT MAY WE NOT EXPECT OF THE NATIVE BORN DAUGHTERS OF [UTAH]? . . . THEY . . . WILL REAR A SUPERSTRUCTURE THAT THE WORLD WILL RECOGNIZE."

—DR. MARTHA HUGHES CANNON

1891

1892

"EQUAL SUFFRAGE IS RAPIDLY GAINING GROUND AND THE RIGHT WILL PREVAIL."

—EMILY S. RICHARDS AND ELECTA BULLOCK

Emmeline B. Wells was elected president of the UWSA on October 5, with Emily S. Richards as vice president. Retiring president Sarah M. Kimball was made honorary president of the association for life.

After a campaign led by Carrie Chapman Catt, Colorado became the second state in the nation to grant full women's suffrage through a narrow victory in a November 7 state referendum.

"WE WHO HAVE ACCEPTED THE NEW GOSPEL OF EQUAL RIGHTS, MUST LABOR WITH UNTIRING ZEAL FOR THE REDEMPTION OF THE MASSES."

—ALVIRA LUCY COX, SANPETE COUNTY SUFFRAGE ASSOCIATION PRESIDENT

1893 | 1894 | 1895

1894

As president of the UWSA, Emmeline B. Wells called a large suffrage meeting in Salt Lake City on February 3. Prominent Utahns such as Dr. Martha Hughes Cannon, Zina D. H. Young, and Justice John E. Booth gave speeches in favor of Utah statehood with equal suffrage rights for women.

Congress passed the Enabling Act on July 16, paving the way for Utah to apply for statehood for the seventh and final time.

Susan B. Anthony wrote to Utah women on July 21, warning them to not allow the women's suffrage question to be separated from Utah's proposed constitution.

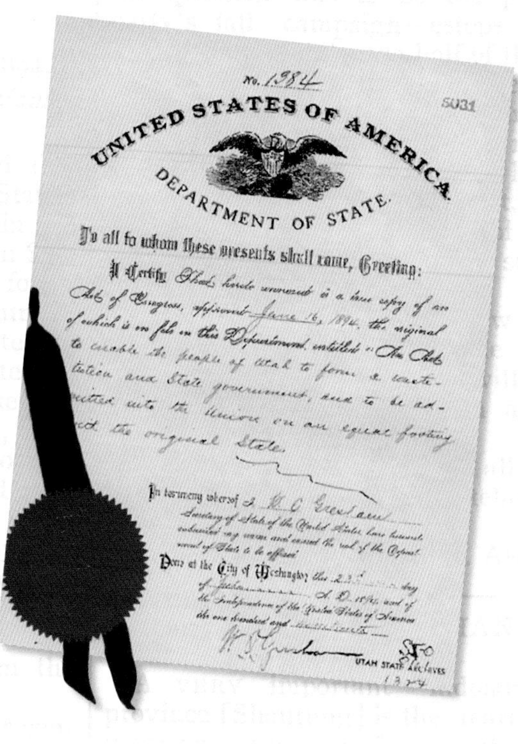

No. 1384

5031

UNITED STATES OF AMERICA.

DEPARTMENT OF STATE.

To all to whom these presents shall come, Greeting:

I Certify That hereto annexed is a true copy of an Act of Congress, approved July 16, 1894, the original of which is on file in this Department, entitled "An Act to enable the people of Utah to form a constitution and State government, and to be admitted into the Union on an equal footing with the original States."

In testimony whereof I, W. Q. Gresham, Secretary of State of the United States, have hereunto subscribed my name and caused the seal of the Department of State to be affixed.

Done at the City of Washington this 23rd day of July A.D. 1894, and of the Independence of the United States of America the one hundred and nineteenth.

W. Q. Gresham

UTAH STATE ARCHIVES

"NOW IN THE FORMATIVE PERIOD OF YOUR CONSTITUTION IS THE TIME TO ESTABLISH JUSTICE AND EQUALITY TO ALL THE PEOPLE. THAT ADJECTIVE 'MALE' ONCE ADMITTED INTO YOUR ORGANIC LAW, WILL REMAIN THERE. DON'T BE CAJOLED INTO BELIEVING OTHERWISE! . . . ONCE IGNORED IN YOUR CONSTITUTION—YOU'LL BE AS POWERLESS TO SECURE RECOGNITION AS ARE WE IN THE OLDER STATES. . . . WITH BEST LOVE TO EACH AND ALL OF YOU—AND BEST HOPE FOR YOUR STATEHOOD."

—SUSAN B. ANTHONY

1890 1891 1892

EQUAL RIGHTS

The women of the Beaver County Woman Suffrage Association drove a "Suffrage carriage" decorated with equal rights banners in Beaver's Pioneer Day parade on July 24. Beaver County had one of the most active suffrage organizations in Utah.

At the Republican territorial convention in Provo on September 11, delegates included a plank in the party platform supporting women's suffrage.

Democratic Party delegates also included support for women's suffrage in their platform at the territorial convention in Salt Lake City on September 15. Electa Bullock, president of the Utah County Woman Suffrage Association, addressed the convention and thanked the party.

"THE DEMOCRATS OF UTAH ARE UNEQUIVOCALLY IN FAVOR OF WOMAN SUFFRAGE, AND THE POLITICAL RIGHTS AND PRIVILEGES OF WOMEN EQUAL WITH THOSE OF MEN INCLUDING ELIGIBILITY TO OFFICE, AND WE DEMAND THAT SUCH GUARANTEES SHALL BE PROVIDED IN THE CONSTITUTION OF THE STATE OF UTAH AS WILL SECURE TO THE WOMEN OF UTAH THESE INESTIMABLE RIGHTS."

—SALT LAKE HERALD-REPUBLICAN, *SEPTEMBER 16, 1894*

1894

Delegates for the constitutional convention were elected on November 6. Local UWSA members had campaigned for pro-suffrage candidates by speaking at political events and holding debates.

—

Speaking at the Salt Lake County Woman Suffrage Association on November 20, president Dr. Ellen Ferguson urged suffragists to visit each of the men elected as constitutional convention delegates. She warned: “Many are inclined to hang back, saying wait till we are a State, then we will give to women suffrage.”

1895

Emmeline B. Wells, Aurelia S. Rogers, and Marilla M. Daniels represented Utah at NAWSA’s annual convention in Atlanta January 31 through February 3. Wells sat next to Susan B. Anthony and spoke on why Utah women should regain the vote. Anthony embraced her and endorsed her message, making “such an eloquent appeal that some of the ladies were moved to tears.”

“SUFFRAGE WOMEN IN THE SEVERAL COUNTIES USED THEIR UTMOST INFLUENCE AND BEST POWERS OF PERSUASION TO DIFFUSE THE IDEAS OF EQUAL RIGHTS.” —*EMMELINE B. WELLS*

Group of State Presidents and Officers of the N.AWSA. at Nat. Convention, 1892.

1890 | 1891 | 1892

By February 1895, there were suffrage organizations in nineteen Utah counties working to educate women in politics and civics, persuade friends and neighbors to support the cause, and lobby constitutional convention delegates.

Utah's constitutional convention opened on March 4 in the new Salt Lake City and County Building. The UWSA had lobbied all 107 male delegates ahead of the convention, but suffrage still became the convention's most hotly debated topic.

The convention first discussed women's suffrage in a March 11 meeting of the Committee on Elections and Suffrage, where delegates approved a suffrage clause taken from the Wyoming state constitution. Some delegates argued that including suffrage might jeopardize congressional approval of the constitution.

1891

Salt Lake County

Woman Suffrage Association

Membersihp Ticket.

Received of ...

Treasurer S. L. C. W. S. A.

"OUR GREAT CREATOR HAS ENDOWED WOMEN WITH THE SAME ATTRIBUTES AND POWER OF MIND POSSESSED BY MAN, HENCE, HER RIGHT TO EQUAL PRIVILEGES AND OPPORTUNITIES UNDER THE LAWS AND IN THE GOVERNMENTS OF THE NATION OF WHICH SHE IS A MEMBER."

—OGDEN AND WEBER COUNTY WOMEN'S PETITION

1893 | 1894 | 1895

1895

On March 18, almost one hundred women from the UWSA met in the Salt Lake City and County Building, just down the hall from the constitutional convention. After drafting a suffrage petition, they filed into the convention hall to hear it read to the delegates.

"THE WOMEN OF UTAH ARE BY NO MEANS INDIFFERENT SPECTATORS OF THE DRAMA THAT IS NOW BEING ENACTED."
—UTAH WOMAN SUFFRAGE ASSOCIATION

That same day, the Salt Lake City Woman Suffrage Association also presented a petition to the convention asking "that the pledges made to the women of Utah be kept and that the new state of Utah may have a Constitution framed upon the basis of justice and privileges for all citizens." Now the women had to watch and wait.

WOMAN SUFFRAGISTS.

A Meeting of the Territorial Association Yesterday.

THE RESOLUTIONS ADOPTED

AND AFTERWARDS PRESENTED TO THE CONVENTION.

A Strong Plea For Equal Suffrage—Women Are a Part of the People—What the Enfranchisement of the Sex Will Do—List of Those Who Signed the Memorial.

The territorial association of woman suffragists met yesterday afternoon at the city and county building and perfected the plan for their campaign in the constitutional convention. The association consists of the leading woman suffragists from all over the territory and their action has behind it the thousands of women who have banded themselves together in county and smaller organizations. As a result

The next morning, on March 19, several women spoke to the Utah Committee on Elections and Suffrage in support of their petitions for suffrage. The speakers included UWSA leaders, Civil War nurse Joanna Melton, and Utah's first woman lawyer, Georgiana Snow Carleton.

1890 1891 1892

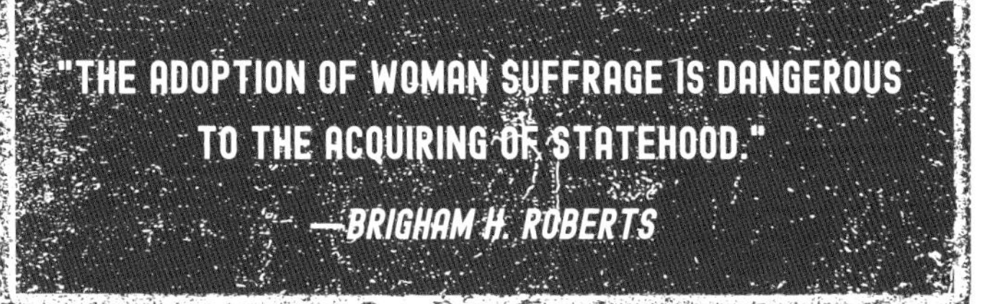

Beginning on March 28, Davis County delegate Brigham H. Roberts unexpectedly argued against enfranchising women in the constitution, claiming it might attract opposition and endanger statehood. His speeches over the next few days breathed new life into anti-suffrage arguments.

At the general Relief Society conference held in the Salt Lake Assembly Hall on April 4, Emily S. Richards urged Latter-day Saint women to continue to support the inclusion of suffrage in the constitution. At the end of her remarks, she asked all in favor of equal suffrage to stand. The conference minutes reported: "Every woman in that large congregation was on her feet immediately."

During the women's meeting, the First Presidency and the Apostles of the Church met in the Salt Lake Temple and discussed women's suffrage at length. With "all committed to woman suffrage," many condemned Brigham H. Roberts and felt that "an enemy could not have betrayed us more." They worried that by stirring up opposition to women's suffrage, Roberts had "aroused latent fears [of Utahns who were not Latter-day Saints] concerning the effect of woman suffrage in giving the Mormons strength."

1893 | 1894 | 1895

1895

Joseph F. Smith spoke powerfully about women's rights at the evening session of the Relief Society conference on April 4, urging that women "not stand in the way of those of their sisters who would be, and of right ought to be free." Apostles Heber J. Grant, Franklin D. Richards, and Charles W. Penrose also spoke in favor of suffrage.

Divisions reemerged among Utah women as they weighed the chance of statehood against their desire for suffrage. Even Charlotte Godbe Kirby publicly supported postponing the suffrage decision.

Women opposed to including equal suffrage in the constitution held a meeting at Salt Lake City's Grand Opera House on April 5. Utah suffragists who were not Latter-day Saints were divided, with several signing a resolution calling to submit the issue to Utah voters as a separate question after statehood (including Jennie Froiseth and Cornelia Paddock), while others (including Emma J. McVicker, Lillie Pardee, and other members of the UWSA) did not sign the resolution.

A WOMAN'S ANSWER.

Mrs. Kirby Favors Postponement Until After Statehood.

Grand Opera House, Salt Lake City, Utah

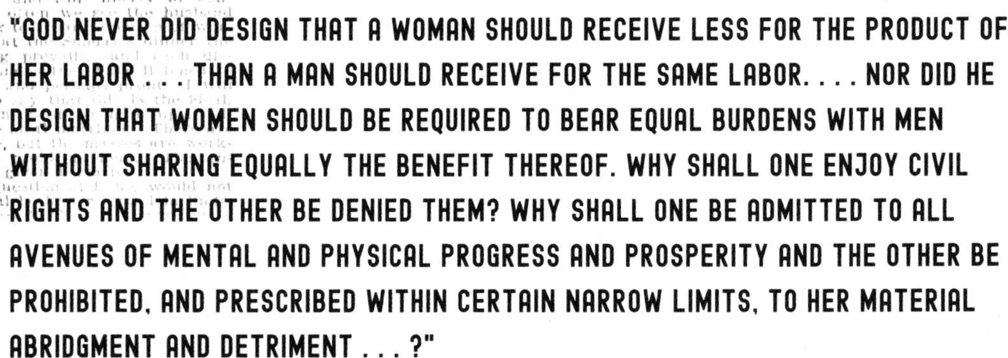

"GOD NEVER DID DESIGN THAT A WOMAN SHOULD RECEIVE LESS FOR THE PRODUCT OF HER LABOR . . . THAN A MAN SHOULD RECEIVE FOR THE SAME LABOR. . . . NOR DID HE DESIGN THAT WOMEN SHOULD BE REQUIRED TO BEAR EQUAL BURDENS WITH MEN WITHOUT SHARING EQUALLY THE BENEFIT THEREOF. WHY SHALL ONE ENJOY CIVIL RIGHTS AND THE OTHER BE DENIED THEM? WHY SHALL ONE BE ADMITTED TO ALL AVENUES OF MENTAL AND PHYSICAL PROGRESS AND PROSPERITY AND THE OTHER BE PROHIBITED, AND PRESCRIBED WITHIN CERTAIN NARROW LIMITS, TO HER MATERIAL ABRIDGMENT AND DETRIMENT . . . ?"

—JOSEPH F. SMITH

1890 1891 1892

Petitions circulated across Utah Territory on both sides of the question. UWSA members such as Mary Ann Freeze gathered signatures on petitions in favor of including equal suffrage in the constitution in Beaver, Box Elder, Cache, Iron, Millard, Salt Lake, Sanpete, Utah, and Wayne counties. Newspapers reported that the final tally was 24,801 signatures for inclusion and 15,366 for separate submission.

On April 5, the convention delegates voted 75 to 14, with 12 absent, to include an equal suffrage clause in the proposed constitution, allowing women to vote and run for political office. This vote was a direct result of Utah suffragists' efforts, and they were optimistic but still cautious. Ruth May Fox noted, "It is expected to be brought up again, so I am afraid the fight is not over."

"[I] COULD AS WELL GO TO THE POLLS TO VOTE AS TO THE POST OFFICE, AND D[O] NOT THINK ANY HOUSES WOULD BE NEGLECTED." —PAULINA LYMAN

***Ruth May Fox (1853–1958)** and her devoted work on behalf of women's suffrage played a key role in restoring women's voting rights in Utah. Fox served as treasurer of the Utah Woman Suffrage Association beginning in 1892. She drafted petitions for equal suffrage and walked all over Salt Lake City soliciting signatures during the 1895 constitutional convention. While lecturing on suffrage in 1895, she declared: "Let us educate [women] until they shall assert their independence and shake the shackles from their wrists, thus may we lift them to a higher plane. . . . I say women to the rescue; women to the front." Despite her undeniable commitment to the suffrage cause and Republican Party, she refused to let herself be dragged into partisan rivalries, declaring: "As for my part I care nothing for politics. It is Mormonism or nothing for me." Fox was a mother of twelve children and later served as president of the Young Ladies' Mutual Improvement Association (YLMIA).*

1893 1894 1895

ORSON F. WHITNEY

"I believe in a future for woman . . . I do not believe that she was made merely for a wife, a mother, a cook, and a housekeeper. These callings, however honorable—and no one doubts that they are so—are not the sum of her capabilities."

"They tell us that woman suffrage in the Constitution will imperil Statehood. I don't believe it. But if it should, what of it? There are some things higher and dearer than Statehood. I would rather stand by my honor, by my principles, than to have Statehood."

"It is woman's destiny to have a voice in the affairs of the government. She was designed for it. She has a right to it. This great social upheaval, this woman's movement that is making itself heard and felt, means something more than that certain women are ambitious to vote and hold office. I regard it as one of the great levers by which the Almighty is lifting up this fallen world, lifting it nearer to the throne of its Creator."

Two of the most outspoken suffrage supporters were Orson F. Whitney and Franklin S. Richards.

FRANKLIN S. RICHARDS

"So I say that if the price of Statehood is the disfranchisement of one-half of the people; if our wives and mothers, our sisters and daughters, are to be accounted either unworthy or incapacitated to exercise the rights and privileges of citizenship, then . . . it is not worth the price demanded."

"Equal Suffrage will prove the brightest and purest ray of Utah's glorious star; it will shine forever in the immortal galaxy, as a beacon light on the tops of the mountains beckoning our sister States and Territories upward and onward to the higher plane of civilization, and the fuller measure of civil and religious liberty."

"I have never known a woman who felt complimented by the statement that she was too good to exercise the same rights and privileges as a man."

Article 4.

Elections and Right of Suffrage.

Section 1. The rights of citizens of the State of Utah to vote and hold office shall not be denied or abridged on account of sex. Both male and female citizens of this State shall enjoy equally all civil, political and religious rights and privileges.

1895

On April 18, the constitutional convention delegates cast the final vote in favor of the women's suffrage clause, ensuring inclusion of women's right to vote and hold office in Utah's proposed state constitution. That constitution was approved by the convention delegates on May 6. Next, it would go to Utah voters for approval.

"THE RIGHTS OF CITIZENS OF THE STATE OF UTAH TO VOTE AND HOLD OFFICE SHALL NOT BE DENIED ON ACCOUNT OF SEX. BOTH MALE AND FEMALE CITIZENS OF THIS STATE SHALL ENJOY EQUALLY ALL CIVIL, POLITICAL AND RELIGIOUS RIGHTS AND PRIVILEGES."

"IT SEEMS ALMOST TOO GOOD TO BE TRUE THAT WE HAVE EQUAL SUFFRAGE."
—EMMELINE B. WELLS

"HURRAH FOR UTAH, NO. 3, STATE."
—SUSAN B. ANTHONY

"WE ALL FELT IT A GREAT DAY IN THE HISTORY OF UTAH."
—RUTH MAY FOX

1893 | 1894 | 1895

1895 *The NAWSA Rocky Mountain Suffrage Convention was held in Salt Lake City May 12–14, bringing suffrage leaders from the western states together with Susan B. Anthony and Dr. Anna Howard Shaw. A procession of seventy-two women met the national leaders at the depot on May 12, and that evening Anthony and Shaw spoke in the Tabernacle to a crowd of more than 6,000. The convention, held at the City and County Building and at the Assembly Hall, included an interdenominational group of participants and demonstrated unprecedented unity among the women of Utah as they celebrated regaining suffrage.*

"I WAS ALWAYS ADVOCATING WOMAN'S RIGHTS AND WOMAN'S ELEVATION FROM MY EARLIEST RECOLLECTION. . . . GOD CREATED US EQUAL, WE STOOD SIDE BY SIDE WHEN MANKIND WAS CREATED."

—MARY ISABELLA HORNE, SALT LAKE STAKE RELIEF SOCIETY PRESIDENT

"THOSE TWO STATES [UTAH AND WYOMING] HAVE THE CREDIT AND WILL HAVE IT THROUGH ALL TIME FOR BEING THE FIRST LEGISLATURES OF THE UNION TO EXTEND FULL SUFFRAGE TO THEIR WOMEN."

—SUSAN B. ANTHONY, NAWSA PRESIDENT

"THE WORK OF THE WORLD DEMANDS THE HIGHEST AND BEST INTERESTS OF MEN AND WOMEN WORKING SIDE BY SIDE TOGETHER."

—DR. ANNA HOWARD SHAW, NAWSA VICE PRESIDENT

"YEARS AGO I WOULD NOT HAVE DARED TO SAY THE BOLD, GRAND THINGS THAT MISS ANTHONY SAID, . . . BUT THE SEED WAS PLANTED WITHIN MY SOUL AND I HAVE BEEN LABORING FOR THE SAME CAUSE."

—SARAH M. KIMBALL, NAWSA HONORARY VICE PRESIDENT FOR UTAH

"WHEN THE LAST BARRIER TO WOMAN IS REMOVED AND THE BALLOT IS PLACED IN HER HANDS THEN WILL SHE SURELY BE A HELPMATE TO MAN."

—LUCY A. CLARK, DAVIS COUNTY WOMAN SUFFRAGE ASSOCIATION PRESIDENT

1895

Approximately one third of the delegates at the Democratic convention on July 13 were women, and several women were appointed to roles in party leadership.

Many women in Salt Lake registered to vote when it appeared that women might be allowed to vote in the fall elections. On August 7, Dr. Martha Hughes Cannon was the first woman to register in Salt Lake City.

It was not clear whether women would be eligible to vote and run for office in the election to ratify the state constitution, so Sarah E. Anderson of Ogden brought a judicial test case on August 10. The Utah Supreme Court eventually ruled that the suffrage clause in the new state constitution would not be valid until after ratification.

The Colored Woman's Republican Club met in Salt Lake City's Independence Hall on August 22. Alice B. Nesbitt, president of the club, chaired the meeting, in which speakers applauded women's suffrage and encouraged black women to register and vote.

IT WAS DEMOCRACY'S DAY

Twelve Hundred Delegates at the Great Convention.

GREAT ENTHUSIASM

General Feeling That the Democratic Tide Has Set In.

WOMEN FULLY RECOGNIZED

They Were Greatly Interested and Quite Prominent.

Mrs. Zina D. Young, Mrs. La Barthe and Mrs. Bullock, of Utah County, Made Officers—Ringing Resolutions Adopted—Thanks to Our Joe—Signal Fires Lighted Anew—Silver to the Front—Able Addresses By Men and Women Leaders—Declaration That Democracy Will Secure the Right of Women to Vote This Fall—Executive Committee Enlarged.

The [illegible] of Utah has kept [illegible] It has furnished to the people of the territory one of the most momentous political conventions within its history. For the first time the women who have helped to make the territory what it is have been admitted into full fellowship with the men in a political gathering on national party lines.

Nor has the experiment proved a failure. Every man in the convention felt that he [illegible] honored by the presence [illegible] many of the women [illegible]

fall, for it will sweep over the whole territory.

Women Recognized.

[illegible] the box reposed the war worn banner of the Tuscaroras. The whole design was neat.

About One-Third Women.

The crowd itself was orderly and [illegible]

RALLY OF COLORED WOMEN.

Enthusiastic Meeting of Second Precinct Voters—Lively Speeches.

The meeting of the Colored Woman's Republican club at Independence hall last night was one of great enthusiasm. The hall was comfortably filled, although the county convention at the Theater proved a strong attraction for those seeking enlightenment on political issues. Mrs. Alice B. Nesbitt, president of the club, acted as chairman, and Lizzie B. Taylor as secretary.

Mrs. I. C. Brown was the first speaker. She reviewed the early history leading up to the forming of the Republican party. The speaker scored the Democratic party on its position on the race question, and its treatment of the colored people of the South, and exhorted the women, who now had been granted the right of franchise, to rise up against the principles of that party, which had fostered slavery and free trade as two of its cardinal principles. She emphasized the necessity for registering, and cautioned those not initiated into the mysteries of Democratic politics to beware of statements made by certain registrars that colored ladies, as well as working girls, were not entiled to register.

John M. Zane made a rousing speech. His replies to many of the objections raised against extending the right of franchise to women caused repeated applause. The real moral force of the country came primarily from woman. As an illustration, he pointed to the abolition movement, which only through the support of woman was carried to a successful termination. He scouted the idea that a woman was too good to go to the polls. "Why," said he, "these women who would have you believe that going to the polls necessarily forces them into contact with bad men, don't have any hesitation in

1890 | 1891 | 1892

The Republican convention on August 23 nominated Emmeline B. Wells for state representative, Lillee Pardee for state senator, and Emma J. McVicker for state superintendent of schools. The three female candidates had to withdraw from the race after the Utah Supreme Court ruled that women could not run or vote in the upcoming election.

UWSA members voted on October 7 to continue meeting even after their own right to vote was reestablished with statehood. The *Woman's Exponent* reported: "All favored the existence of suffrage clubs in the future. They thought it the duty of all to work for the enfranchisement of women until universal suffrage should be obtained."

In the election on November 5, Utah's male voters overwhelmingly approved the proposed state constitution, 28,618 to 2,687.

Utah women sent an onyx and silver ballot box to Elizabeth Cady Stanton for her eightieth birthday celebration in New York City on November 12.

"THERE IS A LARGE MAJORITY FOR STATEHOOD, AND THAT MEANS A LARGE MAJORITY FOR EQUAL SUFFRAGE. THIS IS A MATTER OF REJOICING FOR ALL WOMEN EVERYWHERE WHO HAVE THE ADVANCEMENT OF THE WORLD OF MANKIND AT HEART. IT IS A SINGULAR THING THAT THE WEST . . . SHOULD LEAD."

—WOMAN'S EXPONENT, *NOVEMBER 1, 1895*

1893 1894 1895

UNITED EQUAL SUFFRAGE STATES

OF AMERICA

UTAH

1896

THE THIRD

STATE TO

ENTER

THE UNION OF STATES AS THEY OUGHT TO BE

THIS POSTCARD FEATURING UTAH'S PLACE AS THE THIRD EQUAL SUFFRAGE STATE WAS PART OF THE NATIONAL AMERICAN WOMAN SUFFRAGE ASSOCIATION OFFICIAL POSTCARD SET IN 1910.

SUFFRAGE STATE 1896–1910

UTAH ENTERED THE UNION ON JANUARY 4, 1896, AS THE THIRD state with women's suffrage. The people of Utah celebrated their long-awaited statehood as well as Utah women's regained voting rights. Susan B. Anthony telegraphed: "We all rejoice with you that Utah is a State with her women free and enfranchised citizens." Idaho followed by granting suffrage later that year.

The 1896 general election was the first in which Utah women both voted and ran for office. Seven women ran for state office and voter turnout was high, with women voting in large numbers. Dr. Martha Hughes Cannon was elected as the first female state senator in the nation. Sarah E. Anderson of Weber County and Eurithe LaBarthe of Salt Lake County won elections as state representatives, and eleven other women were elected to county offices.

Utah provided an important example for the rest of the nation of how women's suffrage played out in practice. Anti-suffragists argued that women did not want the vote, that they were incapable of making rational decisions about politics, and that political involvement would distract them from their family duties. Utah women proved otherwise. Election statistics indicated a high level of voter participation among Utah women throughout the first two decades of the twentieth century. In contrast to predictions about harmful outcomes, politically active women influenced Utah families and communities for good.

Meanwhile, suffragists throughout the country were still engaged in a state-by-state campaign while also working toward a constitutional amendment to secure women's suffrage nationally. Utah leaders such as Emmeline B. Wells, Ruth May Fox, Susa Young Gates, and Emily S. Richards continued to foster Utah's relationship with the National American Woman Suffrage Association (NAWSA) and other national women's organizations such as the National Council of Women (of which the Relief Society was a founding member). As the years went by, a new generation of suffrage leaders emerged.

The Utah Council of Women (UCW), founded by Carrie Chapman Catt in 1899 to replace the UWSA, drew women together from multiple faiths and both political parties to help other states in their suffrage efforts. The UCW held monthly meetings and sent delegates to each of the annual NAWSA conventions. Utah women provided funding, served in leadership positions, hosted national leaders, and attended and spoke at national and international women's rights conventions. National suffrage leaders often held up Utah as an example of the success of women's suffrage.

1896

On January 4, President Grover Cleveland signed a proclamation making Utah the forty-fifth state. Because the ratified constitution included women's right to vote, Utah entered the Union as the third equal suffrage state.

Women across Utah celebrated their regained voting rights. Suffrage associations hosted celebrations throughout the state, such as the Cache County Woman Suffrage Association banquet and ball held on January 11 to mark the occasion.

At the NAWSA convention in Washington, D.C., on January 23, national suffrage leaders dedicated an evening to honoring Utah women's achievement. Several Utah women spoke, and a tribute was given to Emmeline B. Wells. Adding a third star to the suffrage flag, Susan B. Anthony said that Utah completed "a trinity of true republics at the summit of the Rockies."

Emily S. Richards, Susan B. Anthony, and other suffrage leaders testified before the U.S. House of Representatives Judiciary Committee in favor of women's suffrage on January 28.

"WOMEN HAVE A CHANCE IN THE UTAH CONSTITUTION TO SHOW THEIR CAPACITY FOR GOVERNMENT, AND HELP MOLD THE INSTITUTIONS OF SOCIETY." —*EMILY S. RICHARDS*

1896 1897 1898

"IF ONE OF YOU MEN DARE TO REFUSE TO RAISE THAT AGE OF CONSENT TO AT LEAST EIGHTEEN YEARS, I WILL SEE THAT YOU NEVER GET ANOTHER WOMAN'S VOTE FOR ANY PUBLIC OFFICE AS LONG AS YOU LIVE IN UTAH!"

—EMMELINE B. WELLS

Political Equality Series.

Vol. IV. No. 6. } NEW YORK, SEPT., 1899. { 10c. per annum. 15c. per hundred

Published Monthly by the National American Woman Suffrage Association, at 107 World Building, New York.

WHERE WOMEN VOTE.

In Utah.

SUSA YOUNG GATES.

I went into the *Woman's Exponent* Office, Salt Lake City, one day soon after Utah women were enfranchised, to speak to the Editor, Mrs. Emmeline B. Wells, who is well known and well beloved in Utah.

"Can you come with me," she said, the moment I entered, "down to the Legislature to appear before a Committee?"

"What about?" I asked.

"Well," she answered, in her nervous, energetic fashion, "the age of consent in this State is now fourteen years; when we settled here and for years after, we had no such law, and indeed no need for it; for you know very well there were almost no cases of crime in our midst. Later years, however, and the influx of civilization and crime, which so often go hand in hand, compelled

Emmeline B. Wells drew on the influence of women's votes while lobbying the Utah legislature in February to raise the age of consent from fourteen to eighteen years old. She threatened reluctant members of the House Legislative Committee with the loss of women's votes if they did not support the bill. After the bill passed, she observed: "It was because we have the franchise and the men know it."

On June 6, Dr. Ellen Ferguson, Dr. Martha Hughes Cannon, Emily S. Richards, and Amanda Knight were elected alternate delegates to the national Democratic convention. Ferguson attended the convention in Chicago as an alternate and urged the committee on resolutions to include a platform plank in favor of women's suffrage.

Suffragist Kate S. Hilliard served as a delegate to the Populist Party's national convention in St. Louis held July 24–26, making her the first Utah woman delegate to a national political party convention.

A nonpartisan women's club worked to register women to vote. Some women reported canvasing three blocks in Salt Lake City with the following result: "Eighty-one for Bryan, twelve for [William] McKinley, while four ladies told them to go home and cook their husbands' dinners."

1899 | 1900 | 1901

Dr. Martha Hughes Cannon (1857–1932) made history as the first female state senator in the United States and made headlines by running against and beating her own husband in the election. She was a doctor, a plural wife, and a devoted suffragist who earned four degrees in medicine and public speaking, ran a private medical practice in Salt Lake City, and was the resident physician of the woman-owned Deseret Hospital. While in office, Dr. Cannon gave birth to her third child, authored Utah sanitation laws, passed several legislative bills that revolutionized public health in the state, and was a founding member of Utah's first State Board of Health. She spoke to U.S. congressional committees as well as national suffrage conventions in favor of women's suffrage. Dr. Cannon argued that "one of the principal reasons why women should vote—is that all men and women are created free and equal."

"YOU GIVE ME A WOMAN WHO THINKS ABOUT SOMETHING BESIDES COOK STOVES AND WASH TUBS AND BABY FLANNELS, AND I'LL SHOW YOU, NINE TIMES OUT OF TEN, A SUCCESSFUL MOTHER." —*DR. MARTHA HUGHES CANNON*

1896

After the Republican-leaning *Salt Lake Tribune* endorsed Angus Cannon as a senate candidate, the *Salt Lake Herald* responded on October 31 by endorsing his wife, who was running on the Democratic ticket: "Mrs. Mattie Hughes Cannon, his wife, is the better man of the two. Send Mrs. Cannon to the state senate as a Democrat and let Mr. Cannon, as a Republican, remain at home to manage home industry."

In an at-large election on November 3, eight men and two women ran for the five open Utah Senate seats in Salt Lake County, with each gender evenly divided between the Republican and Democratic Parties. Democrat Dr. Martha Hughes Cannon won one of the seats, beating out all the Republicans including her husband and Emmeline B. Wells.

Democrats Eurithe LaBarthe and Sarah E. Anderson also won the November 3 election as state representatives for Salt Lake County and Weber County, while Republicans Lucy A. Clark of Davis County, Martha Campbell of Salt Lake County, and Fanny E. Stewart of Utah County were defeated.

Eleven other women were elected to county offices across Utah on November 3: Margaret A. Caine, Ellen Jakeman, Delilah K. Olson, Lottie Farmer, Maude Layton, Mamie Wooley, Amelia Graehl, Emily Dods, Tryphenia West, Bessie Morehead, and Mary F. Shelby.

Idaho became the fourth state to grant women's suffrage on November 3. The NAWSA *History of Woman Suffrage* credited Utah with helping obtain suffrage in Idaho: "A strong factor in the campaign was the large colony in the Southern part of the State who were residents of Utah when women voted there and who believed in their enfranchisement. Mrs. Emily S. Richards of Utah did effective work among them."

"AND NOW IDAHO COMPLETES THE QUARTETTE OF MOUNTAIN STATES WHICH SING THE ANTHEM OF WOMAN'S FREEDOM."

—CLARA B. COLBY

"FIFTY YEARS AGO WOMEN VOTED NOWHERE IN THE WORLD; TO-DAY WYOMING, COLORADO, UTAH AND IDAHO HAVE ESTABLISHED EQUAL SUFFRAGE FOR WOMEN."

—IDA HUSTED HARPER

1899 | 1900 | 1901

"[UTAH IS] A COMPLETE VINDICATION OF THE EFFORTS OF EQUAL SUFFRAGISTS. . . . NONE OF THE UNPLEASANT RESULTS WHICH WERE PREDICTED HAVE OCCURRED."

—DR. MARTHA HUGHES CANNON

1897

With Utah suffrage secure, the *Woman's Exponent* masthead changed in January to read: "The Ballot in the Hands of the Women of Utah should be a Power to better the Home, the State and the Nation." This reflected Latter-day Saint women's focus on using their political power to improve society.

At a NAWSA convention in Des Moines, Iowa, on January 26, Emmeline B. Wells had the honor of introducing Susan B. Anthony to give the keynote address. Wells and her daughter Melvina Woods, a delegate for the newly-enfranchised women of Idaho, were both invited to sit with Anthony on the stand and to address the convention as well as the Iowa Senate.

1898

State Senator Martha Hughes Cannon addressed the NAWSA convention in Washington, D.C., on February 13. She also testified to the House Judiciary Committee two days later in favor of a federal women's suffrage amendment.

WOMAN'S EXPONENT.

The Ballot in the Hands of the Women of Utah should be a Power to better the Home, the State and the Nation.

VOL. 25. SALT LAKE CITY, UTAH, JAN. 15, AND FEB. 1, 1897. No. 14 & 15

CONTENTS:

A Happy Birthday—L. D. Alder. A Mother In Israel—Maria McRae. U. W. P. C.—Orielle Curtis, Sec. Relief Society In Chicago. Anniversary Party—Mary S. Christensen. Spring City Relief Society—Nancy Acord. Mrs. Stowe's Monument. R. S. Reports. In Memoriam—M. A. Hardy. Resolutions Of Respect. Letter From Columbus, Kansas—E. A. Scammon. Obituary —A. C. M. Ladies' Semi-Monthly Meeting—L. D. Alder Sec. Washington Co. Silk Asso'n. Lula's Letter—L. L. Greene Richards. Notes And News.

EDITORIAL:—Annual Convention N-A.W. S. A. Mary Houston Kimball Smith Re-organization At Brighton. Editorial Notes.

POETRY:—To Dr. Ellis R. Shipp On Her Birthday Jubilee—Emily H. Woodmansee My Only Sister—L. D. Alder. Selection From Times And Seasons—President Joseph Smith.

"MY ONLY SISTER."

In the gladsome Long Ago
In the bright sunshining weather
There romped by my side a winsome lass
As we strolled along together.
The light wind caught her flaxen hair
And smothered the fair young face,
While her eyes of blue flashed in merriment
As she and the wind kept apace.

I smiled at her glee and mischief
She danced so much like an elf;
And though I tried to be serious,
I laughed in spite of myself.
Oh, those happy days are gone forever,
For my sister to womanhood grew,
Crowned with a beauty and gentleness,
That were both strange and new.

But O, I know she is living
In a world where all is love,
And I gaze on the stars in the silent night,
As they shine in the vault above.

And I think of how she is waiting,
And watching for me to come,
And I feel that she knows how I loved her,
While here in this earthly home.
And changes now will never come,
Her heart and mine between;
And never a tear will fall again,
For the love that "might have been."

She is still my little sister
As in the far away days;
And I put out my hand to guard her,
Again in her romping ways.
And I am still her brother

1896 1897 1898

"A LIFE CONSUMED BY FOLLOWING SOCIETY'S UNPROFITABLE AND FOOLISH FASHIONS HAS A PARALLEL IN THAT OF A WOMAN WHO NEVER TAKES A MOMENT FOR STUDY AND SELF-IMPROVEMENT, BUT MAKES HERSELF A VERY SLAVE TO HER HOME. THE HOME MUST BE KEPT SWEET AND CLEAN, BUT THE BRAIN IS AS PRONE TO GET COBWEBBY AS THE BEST ROOM."

—ALICE MERRILL HORNE

ECHOES OF THE ELECTION.

When Mrs. B. B. Nesbitt, W. W. Taylor, H. H. Voss, Mrs. W. W. Taylor and Uncle Andrew Campbell appeared on the streets last Tuesday morning, they were surprised and startled to observe that fifteen or twenty of the best and most influential members of our race were engaged in working in the interest of the Democratic party. It had been reported on

1898

Utah's small population of black men and women voted and actively participated in party politics. Elizabeth Taylor and Alice Nesbitt, leaders of the Colored Woman's Republican Club, vigorously campaigned for the Republican Party in the November election. They were joined by most black voters, although the Democratic-leaning newspaper, *The Broad Ax*, reported that a substantial number of black citizens had voted the Democratic ticket.

On November 8, Democrat Alice Merrill Horne, granddaughter of early suffragist Bathsheba W. Smith, was the third Utah woman elected to the Utah House of Representatives. Her legislative successes included securing the University of Utah land grant, passing a four-year scholarship bill, and establishing the Utah Arts Council as the nation's first state-sponsored art collection.

"EVERY PERSON, BLACK OR WHITE, HAS THE RIGHT TO AFFILIATE WITH ANY POLITICAL PARTY."

—THE BROAD AX, *NOVEMBER 12, 1898*

1899 | 1900 | 1901

National Council of Women
of the United States

February 1, 1899.

Mrs. Susa Young Gates,
Provo City, Utah.

My dear Mrs. Gates:--

Let not your heart be heavy. Yours of January 28th is received. All the errors which you point out are corrected. We must all of us do our "humble, quiet best."

You need not be afraid that you will have difficulties of an unusual character to encounter in Washington. I believe you will find your way paved. Certainly I shall help you all I can to have an opportunity to do the good service in the Council for which you are so well fitted.

Affectionately yours,

May Wright [...]

President of the Nat[...]
of Women of the Uni[...]

P.S.-

[...]n women ought to try
[...]n do nothing about th[...]
[...]ncil work is too gre[...]

UTAH WOMEN AT WASHINGTON

SUCCESSFUL IN THEIR DEFENSE OF UTAH.

Express Their Gratitude to Susan B. Anthony and Belva A. Lockwood —Mrs. Card's Address.

"I . . . SHALL STAND FOR THE RECOGNITION OF MORMON WOMEN ON EXACTLY THE SAME BASIS THAT I WOULD STAND FOR THE RECOGNITION OF MY OWN CHURCH. I WISH THEM [TO] . . . ENTER THE COUNCIL, AND TO HAVE AS CLOSE AN AFFILIATION BETWEEN THEM AND WOMEN OF ALL OTHER DENOMINATIONS AS CAN BE BROUGHT ABOUT."

—MAY WRIGHT SEWALL

1899

Hannah Kaaepa and nine other Utah representatives traveled to Washington, D.C., for the third Triennial Congress of the National Council of Women (NCW) held February 11–20. As a native Hawaiian living in Utah, Kaaepa spoke on behalf of Hawaiian women and advocated for women's suffrage in the new territory of Hawaii. Emmeline B. Wells was elected as the NCW Assistant Recording Secretary, and Susa Young Gates served on the Program Committee and as the temporary Chairman of the Press Committee.

Controversy arose at the NCW conference when a proposed resolution denounced the election of Brigham H. Roberts to Congress due to his practice of plural marriage. The Relief Society delegates were placed in the difficult position of having to choose between defending Roberts, a previous suffrage opponent, or taking NCW President May Wright Sewall's advice to use the "golden opportunity" to demonstrate unity with other women's organizations. Latter-day Saint delegates maintained their loyalty to the Church and defeated the resolution.

1897 1898 1899

As a medical doctor, state senator Martha Hughes Cannon was deeply committed to securing passage of her Public Health Bill, which provided standards and regulations for disease prevention. She introduced the bill to the state Senate on February 14, and it passed a few weeks later.

Even before declaring her candidacy for the state legislature, Alice Merrill Horne drafted the bill that ultimately created the Utah Art Institute and the state's "Alice Art Collection." The act passed the Utah House and Senate on March 9. As Utah Governor Heber M. Wells signed the Art Bill into law, he declared the bill "the direct result of equal suffrage."

As the only two women serving in Utah's 1899–1900 legislative session, Senator Cannon and Representative Horne joined forces to achieve their most important legislative victories. During debates over both Cannon's Public Health Bill and Horne's Art Institute Bill, they scattered yellow flowers on the desks of male legislators. This symbol of suffrage reminded the men of the influence the female legislators wielded among female voters.

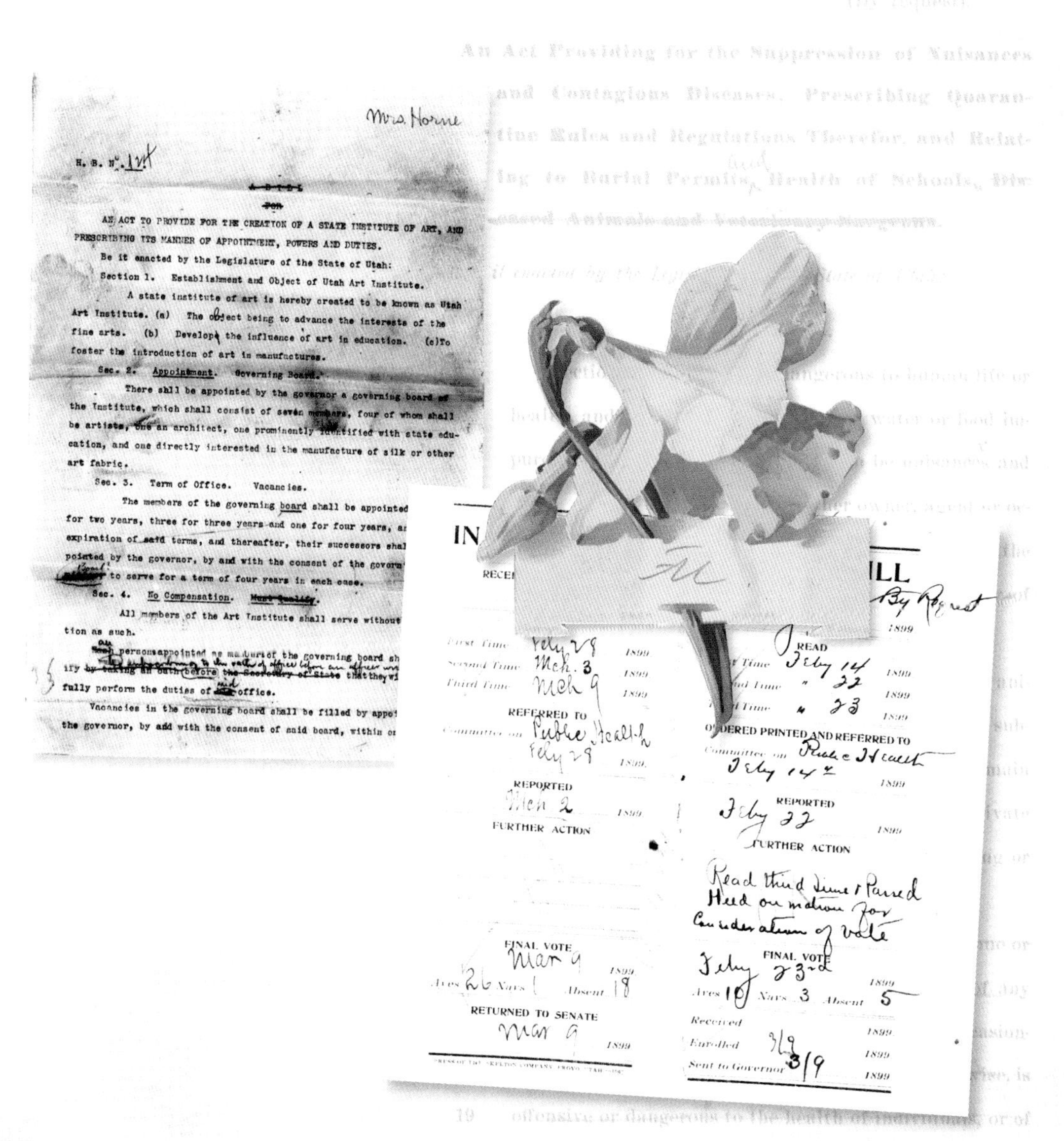

S. B. No. 40. By Mrs. CANNON. (By request)

An Act Providing for the Suppression of Nuisances and Contagious Diseases, Prescribing Quarantine Rules and Regulations Therefor, and Relating to Burial Permits, Health of Schools, ~~Diseased Animals and Veterinary Surgeons.~~

Mrs. Horne

H. B. No. 128

~~A BILL~~

~~FOR~~

AN ACT TO PROVIDE FOR THE CREATION OF A STATE INSTITUTE OF ART, AND PRESCRIBING ITS MANNER OF APPOINTMENT, POWERS AND DUTIES.

Be it enacted by the Legislature of the State of Utah:

Section 1. Establishment and Object of Utah Art Institute.

A state institute of art is hereby created to be known as Utah Art Institute. (a) The object being to advance the interests of the fine arts. (b) Develop the influence of art in education. (c) To foster the introduction of art in manufactures.

Sec. 2. Appointment. Governing Board.

There shall be appointed by the governor a governing board of the Institute, which shall consist of seven members, four of whom shall be artists, one an architect, one prominently identified with state education, and one directly interested in the manufacture of silk or other art fabric.

Sec. 3. Term of Office. Vacancies.

The members of the governing board shall be appointed for two years, three for three years and one for four years, a[t the] expiration of said terms, and thereafter, their successors shal[l be ap]pointed by the governor, by and with the consent of the govern[ing] board to serve for a term of four years in each case.

Sec. 4. No Compensation. ~~Must Qualify~~.

All members of the Art Institute shall serve without [compensa]tion as such.

All persons appointed as members of the governing board sh[all qual]ify ~~by taking an oath before the Secretary of State~~ that they wi[ll faith]fully perform the duties of said office.

Vacancies in the governing board shall be filled by appo[intment of] the governor, by and with the consent of said board, within o[...]

IN[TRODUCED] ... [B]ILL

By Request

First Time	Feby 28 1899	First Time	Feby 14 1899
Second Time	Mch. 3 1899	Second Time	" 22 1899
Third Time	Mch 9 1899	Third Time	" 23 1899

REFERRED TO Committee on Public Health Feby 28 1899

ORDERED PRINTED AND REFERRED TO Committee on Public Health Feby 14 1899

REPORTED Mch 2 1899

REPORTED Feby 22 1899

FURTHER ACTION

FURTHER ACTION Read third time / Passed. Held on motion for Consideration of vote

FINAL VOTE Mar 9 1899 Ayes 26 Nays 1 Absent 18

FINAL VOTE Feby 23rd 1899 Ayes 10 Nays 3 Absent 5

RETURNED TO SENATE Mar 9 1899

Received 1899

Enrolled 3/9 1899

Sent to Governor 3/9 1899

Carrie Chapman Catt (1859–1947) was elected president of NAWSA in 1900, just months after her trip to Utah. She established the International Woman Suffrage Alliance in 1902 and served as the Alliance's president even after stepping down from NAWSA in 1904 to care for her dying husband. Catt resumed leadership of NAWSA from 1915 to 1920 and devised the "Winning Plan," which coordinated the state suffrage campaigns with a careful drive for a constitutional amendment. Just before passage of the Nineteenth Amendment, she founded the national League of Women Voters and returned to Utah once more to establish the Utah League of Women Voters.

"NO GREATER EVIDENCE OF THE ILLIMITABLE SCOPE OF THE SO-CALLED WOMAN'S MOVEMENT COULD BE FOUND THAN WAS GIVEN IN THE GREAT CONGRESS RECENTLY HELD IN LONDON."
—*SUSA YOUNG GATES*

MRS. CARRIE CHAPMAN CATT and Miss Mary G. Hay, of New York City, are touring in the West in the interest of Equal Suffrage. Both are fine looking, and are women of marked ability. Mrs. Catt is one of the finest speakers on the platform, that there is in the suffrage work. They came here from Montana and went direct from Salt Lake to San Francisco. At a meeting of ladies in this office on Monday, October 30, at 3 p. m., a Central Committee was organized to do special work in this State in the interest of suffrage for other parts of the country. Mrs. F. S. Richards is chairman; Mrs. J. Fewson, Smith secretary; Mrs. Ella W. Hyde, treasurer; Mrs. E. J. McFarlane, chairman canvassing committee. The ladies represented three counties. A meeting to complete the organization and plan will be held at Mrs. P. P. Jennings' residence, on Saturday at 3 p. m., November 4.

DELEGATES TO GO TO LONDON

MEETING OF THE INTERNATIONAL COUNCIL OF WOMEN.

Mrs. Emeline B. Wells of Salt Lake Among the Women Leaders From This Country.

New York, June 6.—The departure of Mrs. May Wright Sewall of Indiana on the Bremen Thursday marks the exodus of American women to London for the quinquiennial meeting of the International Council of Women, which will open Monday, June 26, and continue until Wednesday, July 5, inclusive. At present the vice president and prospective president of the International Council, Mrs. Sewall, naturally leads the 100 American women who will attend and in most cases participate in this convention of world-wide import.

1899

Susa Young Gates and Emmeline B. Wells were among the sixteen Utah women who attended the International Council of Women conference in London from June 26 to July 5, which focused on women's suffrage and international peace.

Carrie Chapman Catt visited Utah and met with twenty-five suffragists in the office of the *Woman's Exponent* on October 30. They formed the Utah Council of Women (UCW), a Utah branch of NAWSA, with Emily S. Richards as president and Emmeline B. Wells on the national executive committee. The UCW replaced the UWSA and included Utah women of multiple faiths and political parties.

1898 1899 1900

"MY PLEASURE IN THE RICH BROCADED SILK IS QUADRUPLED BECAUSE IT WAS MADE BY WOMEN POLITICALLY EQUAL TO MEN."

—SUSAN B. ANTHONY

"WE RESENT THE INTIMATION THAT IN SPITE OF EQUAL SUFFRAGE THE WOMEN OF UTAH REMAIN POLITICAL NONENTITIES. . . . WHERE LABOR AND RESPONSIBILITY ARE EQUAL WE CONTEND FOR EQUAL COMPENSATION."

—LADIES' DEMOCRATIC CLUB OF PROVO

1900

Suffragists celebrated Susan B. Anthony's eightieth birthday at the NAWSA convention in Washington, D.C., on February 15. Utah delegates Lucy A. Clark and Emily S. Richards presented Anthony with a bolt of black silk, produced entirely by Utah women, from which Anthony had her cherished "Utah dress" made. The dress is now on display at the Susan B. Anthony House.

Elizabeth Cohen of Salt Lake City served as a delegate to the Democratic National Convention in Kansas City in July, where she seconded the presidential nomination of William Jennings Bryan. Jennie Jones, president of the Salt Lake City Woman's Republican Club, was an alternate delegate to the Republican National Convention in June.

The Ladies' Democratic Club of Provo, led by Amanda Knight, protested the Democratic Party's plan to nominate only men to the county ticket for the upcoming election. At a meeting on August 23, the women also objected to salary discrimination against female county employees.

1901 | 1902 | 1903

Emma J. McVicker (1846–1916) was Utah's first female state superintendent of schools and one of the few women actively involved in the Utah Woman Suffrage Association who was not a Latter-day Saint. A prominent suffragist, she helped host Susan B. Anthony and Anna Howard Shaw during their visit to Salt Lake City in 1895 and joined them and other Utah suffrage leaders on the stand in the Tabernacle during the Rocky Mountain NAWSA convention. McVicker established Utah's first free public kindergarten and worked as principal and music teacher at the Salt Lake Collegiate Institute (later Westminster College). Later she was elected as the first female regent for the University of Utah. McVicker earned a master's degree from the University of California at age fifty-nine and started a fund to provide no-interest loans to young women who were working their way through university.

1900

Lucretia Boynton organized women's Republican clubs in eight Cache County towns and was the only woman to speak at the state Republican convention in September.

Emma J. McVicker became the first woman named to a post in Utah state government when she was appointed state superintendent of schools on October 8 to fill a three-month vacancy. She had been nominated for the same position by the Republican Party in 1895 but was prevented from running prior to the adoption of the state constitution.

President Lorenzo Snow, Joseph F. Smith, Heber J. Grant, and other Church leaders blessed Emmeline B. Wells on November 9 prior to her departure for an NCW executive committee meeting. They set her apart and blessed her with "influence with the women among whom she may associate in this Convention, and grant that she may have power with them, . . . that the rights and privileges which belong to the women of Thy people in the midst of these mountains may be recognized and acknowledged by the women of the nation and by all the people of the nation."

1899 1900 1901

The UCW donated items for NAWSA's fundraising bazaar held December 3–8, including a quilt made from Utah silk and a doll donated by the Utah governor. The four enfranchised states had a prominent booth next to the main platform.

1902

Emmeline B. Wells attended what would be her last national suffrage convention February 12–17 in Washington, D.C., in conjunction with the first International Woman Suffrage conference. Wells sat on the platform in the opening session next to Carrie Chapman Catt, May Wright Sewall, Clara Barton, Susan B. Anthony, and Anna Howard Shaw.

1902 | 1903 | 1904

1903

Utah state representative Mary G. Coulter of Ogden was the first woman in the nation to chair a state House Judiciary Committee. Coulter, one of Utah's early female lawyers, declared: "I have always been for equal rights for men and women. My mother was a strong suffragist."

MRS. MARY G. COULTER.

THE OFFICERS, DELEGATES, AND SPEAKERS OF THE I. C. W. ASSEMBLED IN THE BANQUET ROOM OF THE PALAST HOTEL, BERLIN, AT A LUNCHEON GIVEN BY MRS. SEWALL, THURSDAY, JUNE 9, 1904

1904

Corrine Allen, Alice Merrill Horne, Lydia Alder, Emily S. Richards, and Ida Dusenberry represented Utah organizations at the International Council of Women Congress held in Berlin June 8–18. The Congress established the International Suffrage Alliance, with Carrie Chapman Catt as president. Fueled by renewed anti-polygamy attacks by fellow Utah suffragist Corrine Allen, tensions once again escalated over the membership of Latter-day Saint women in the council. Ultimately, their tenuous but carefully cultivated relationships with national suffrage leaders prevailed, and Latter-day Saint organizations continued active membership in the ICW and NCW.

Mrs. Lydia D. Alder, Salt Lake City, Utah.

Mrs. Clarence E. Allen, Salt Lake City, Utah.

"THE WOMEN OF UTAH WHO ENJOY THE BOON OF SUFFRAGE, DO NOT FORGET THAT IT IS THEIR DUTY TO DO ALL THEY CAN, THAT THE SISTERS IN OTHER STATES MAY RECEIVE THE POWER AND OPPORTUNITY OF DOING GOOD THROUGH THE BALLOT."

—DESERET EVENING NEWS, *APRIL 6, 1904*

1901 1902 1903

1904

Delegates from seven western states attended a convention organized by Elizabeth Taylor in Salt Lake City July 5–7 to organize the Western Federation of Colored Women. Taylor was elected president of the federation and editor of the monthly newspaper, *The Western Women's Advocate*. Convention delegates visited Saltair and were entertained by Salt Lake's Colored Women's Progressive Club.

1905

The Western Federation of Colored Women held a large fund-raiser in Salt Lake City to establish an orphanage and a home for the elderly. The work of the organization reflected the motto of the National Association of Colored Women, "Lifting as We Climb," by advocating for black women while also working to improve the status of all black Americans.

"I BELIEVE THAT THE COLORED WOMEN SHOULD STAND TOGETHER MORE THAN ANY OTHER CLASS OF CIVILIZED WOMEN IN THE WORLD."
—ELIZABETH TAYLOR

MRS. W. W. TAYLOR,

Black women's advocates throughout the nation faced additional restrictions on their rights and opportunities because of their race. Women's rights organizations throughout the nation generally excluded black women, and white suffragists did not often address racial discrimination in their campaigns. In response, black women in many parts of the country established their own organizations to advance their rights. The National Association of Colored Women (NACW) was founded in 1896 as a federation of these local clubs. While suffrage was an important goal for black women reformers, they worked for a broader range of reforms to make life better for all black Americans. Across the country, they fought lynching, segregation, and Jim Crow laws instituted to keep black men from exercising their voting rights. They also worked for better educational opportunities for black men, women, and children. In Utah, black women and men were active political participants and voters. Black women in Utah founded several women's clubs to organize charitable work and social opportunities.

1904 1905 1906

DESERET EVENING NEWS.

TRUTH AND LIBERTY.

32 PAGES—LAST EDI

HONOR LIFE OF MISS ANTHONY.

Deeply Impressive Services in Memory of "Foremost American Woman."

TELL OF GREAT LIFE WORK.

Resolutions of Respect and Expressions of Love Heard At Today's Meeting.

—Week's Cleverest Cartoon, by Triggs, in the New York Press.

WOMEN BESIEGE NATIONAL CAPITOL, SEEKING SUFFRAGE

Present Arguments Before Senate and House Committees to Show Why They Should Be Allowed to Vote.

Makes Stirring Speech In Eulogy of the Sex—Their Influence, He Declares, to Balance the Saloon Element, Which Caused the Defeat of Suffrage in Oklahoma.

C.—Almost 100 the Capitol in They argued before committee on Judescended on the on Woman Sufhours they talked. shooed from the the Senate by the Mrs. Harriet Taylor Upton, of Ohio. Among the speakers were Miss Emma Gillette, of Washington; Mrs. Catt, Mrs. Richard W. Fitzgerald, of Massachusetts; Senator Owen, Miss Rose Sullivan, of Utah; Mrs. Mary E. Craigie, of New York; Mrs. Ida Porter Boyer, of Pennsylvania; and Miss Gordon, of Louisiana.

UTAH WOMAN PLEADS FOR WOMEN SUFFRAGISTS

WASHINGTON, D. C., March 3.—The advocates of female suffrage were today given their annual opportunity to present pleas to Congress, the presentation to the Senate being made before the committee on woman suffrage, and to the House before the judiciary committee. Before the Senate committee meeting, Rev. Anna Shaw, as president of the National Female Suffrage association, introduced the speakers, the first of whom was Mrs. Belva Lockwood, who expressed confidence in the support of her cause by the committee. Mrs. Fannie Fernald made an eloquent plea for "a voice in the government which controls every interest we hold dear." Mrs. Harriet Taylor Upton of Ohio was in charge of the House committee, and the speakers included Miss Emma Gillette of Washington, Mrs. Chapman Catt of New York, Senator Owen of Oklahoma, and Miss Rose Sullivan of Utah.

1906

Upon the death of Susan B. Anthony on March 13, the UCW gathered for a memorial service in the Salt Lake City Assembly Hall. Emily S. Richards, Emmeline B. Wells, Ruth May Fox, and Alice Merrill Horne commemorated Anthony's life and urged continuing commitment to the suffrage cause that she held so dear. Anthony was buried with a jeweled suffrage flag bearing a star for each of the four suffrage states, including Utah. She had bequeathed a gold ring to Emmeline B. Wells as a token of their friendship.

1908

Utah suffragist Rose Sullivan testified in favor of women's suffrage alongside Carrie Chapman Catt and other NAWSA leaders before the U.S. House of Representatives Judiciary Committee on March 3.

1904 1905 1906

"WE WOMEN OF UTAH RIDE TO THE POLLS JUST AS HARMLESSLY AS WE RIDE TO CHURCH." —*LUCY A. CLARK*

Lucy A. Clark, an "ardent woman suffragist," earned national distinction as the first woman to vote as a delegate and give a speech at a Republican National Convention in June. While in Chicago, Clark also spoke at a women's club gathering, stating, "We are asking for a serious consideration of suffrage as a subject that intimately concerns the welfare of the nation."

Elizabeth Hayward was elected as a Utah delegate to the National Democratic Convention in Colorado in July. She then accompanied Lucy A. Clark to a NAWSA convention in Buffalo, New York.

"MRS. ELIZABETH HAYWARD OF SALT LAKE CITY, WHO . . . SERVED AS DELEGATE TO THE DEMOCRATIC NATIONAL CONVENTION IN DENVER, IS THE MOTHER OF NINE CHILDREN, AND SAID TO HAVE ONE OF THE BEST MANAGED HOUSEHOLDS IN THAT CITY. SHE IS AN ARDENT ADVOCATE OF EQUAL SUFFRAGE, NOT BECAUSE IT 'BROADENS' THE VIEWS OF THE MODERN WOMAN, BUT BECAUSE IT GIVES THEM THE POWER TO PROTECT AND IMPROVE THEIR HOMES."

—SALT LAKE TELEGRAM, *SEPTEMBER 23, 1908*

1907 | 1908 | 1909

1909

The NAWSA fundraising bazaar in New York City held December 10–11 featured dolls from the four original suffrage states. Carrie Chapman Catt thanked Sarah Elizabeth Cutler, wife of the former Utah governor, for donating a "little voter from Utah" doll.

1910

The National American Woman Suffrage Association sold suffrage stamps featuring Utah among the four states in which women enjoyed equal suffrage rights. U.S. President William Howard Taft purchased the first suffrage stamp on April 10 as a show of support for the cause.

The UCW collected 40,000 names in Utah for a NAWSA petition seeking a national suffrage amendment. The full petition, with 400,000 names, was presented to Congress on April 19 in a mile-long procession of decorated cars carrying the signatures.

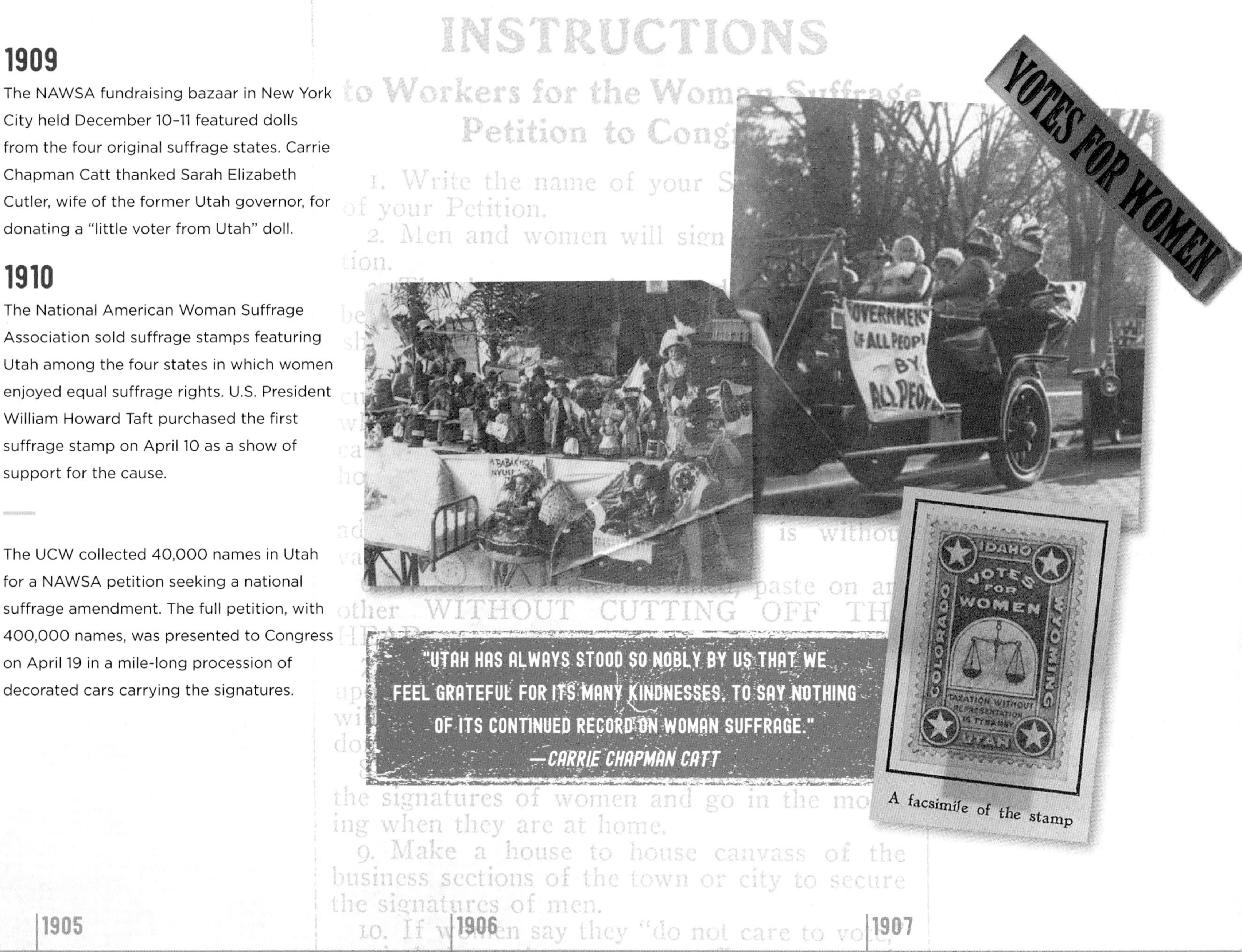

"UTAH HAS ALWAYS STOOD SO NOBLY BY US THAT WE FEEL GRATEFUL FOR ITS MANY KINDNESSES, TO SAY NOTHING OF ITS CONTINUED RECORD ON WOMAN SUFFRAGE."
—CARRIE CHAPMAN CATT

A facsimile of the stamp

1905 1906 1907

Emmeline B. Wells was appointed to serve as the General President of the Relief Society on October 3, making her the leader of over 33,000 Latter-day Saint women at the age of eighty-two. Wells used the relationships she had developed through her suffrage work to continue building bridges between women within the Church and those of different faiths.

On November 8, Washington state's male voters approved women's suffrage, breaking a fourteen-year gridlock in the suffrage movement. Washington women had previously gained the right to vote in 1883, but the territorial Supreme Court had twice overturned the law on a technicality. This victory revitalized the national suffrage movement.

UNCLE SAM'S NEWEST GIRL BABY---HE HAS FIVE DAUGHTERS NOW

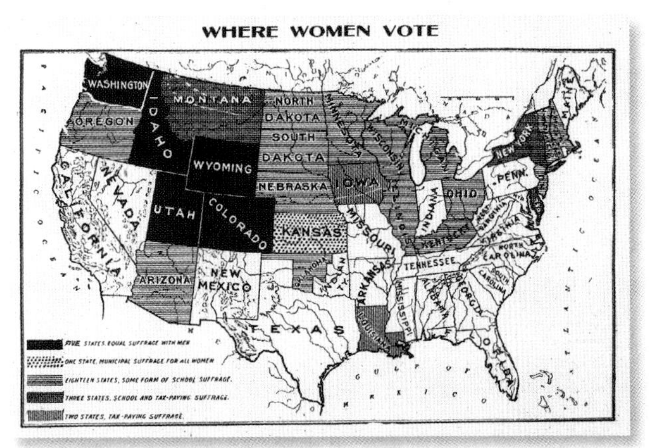

"THIS VICTORY, SO GRATIFYING TO THE WOMEN OF WASHINGTON, HAD ALSO AN IMPORTANT NATIONAL ASPECT, AS IT MARKED THE END OF THE DREARY PERIOD OF FOURTEEN YEARS FOLLOWING THE UTAH AND IDAHO AMENDMENTS IN 1895–6, DURING WHICH NO STATE ACHIEVED WOMAN SUFFRAGE."

—IDA HUSTED HARPER

1908 | 1909 | 1910

UTAH SUFFRAGISTS AND LEADERS FROM THE CONGRESSIONAL UNION MET WITH UTAH SENATOR REED SMOOT IN AUGUST 1915 TO URGE HIS CONTINUED SUPPORT FOR A FEDERAL AMENDMENT FOR WOMEN'S SUFFRAGE.

6 SECURING NATIONAL SUFFRAGE 1911–1920

MORE THAN SIXTY YEARS AFTER THE WOMEN'S MOVEMENT BEGAN in the United States, Utah was still one of only five states that recognized equal suffrage for women. Just a decade later, in 1920, a constitutional amendment extended suffrage to women across the nation. The continued involvement of many Utah women in the suffrage movement, even after they regained their own right to vote, demonstrates their commitment to the cause of women's suffrage itself.

As a new generation of women joined the suffrage cause, disagreements over tactics and philosophy once again split the national movement. After organizing an influential NAWSA parade, Alice Paul and Lucy Burns broke away from NAWSA in 1913 and formed the more progressive Congressional Union for Woman's Suffrage (CU), which became the National Woman's Party (NWP) in 1916. Although Utah suffragists primarily maintained their allegiance to NAWSA through the Utah Council of Women (UCW), led by Emily S. Richards, many also supported the efforts of the CU and to some extent the NWP. They held rallies, attended conventions, provided funds, and gathered signatures on petitions for both NAWSA and the CU. Utah Senator George Sutherland introduced the "Susan B. Anthony Amendment" to the U.S. Senate in December 1915.

While most Utah suffragists did not support the NWP's increasingly militant tactics, two Utah women were arrested with other NWP members while picketing in front of the White House in 1917. Highly publicized reports of the prisoners' mistreatment generated public sympathy for suffragists. Once President Wilson announced his support for a suffrage amendment, Congress passed the Nineteenth Amendment in the spring of 1919. Utah ratified the amendment in a special legislative session that October, with female legislators playing key roles. Utah suffragists continued to meet as the League of Women Voters.

In August 1920, the Nineteenth Amendment to the U.S. Constitution became law and extended women's voting rights throughout the United States. Utah women celebrated this important step toward women's equality in the same year they celebrated the fiftieth anniversary of their own suffrage.

Susa Young Gates (1856–1933) cast her first vote as a young newlywed in 1872. A daughter of Brigham Young, she was a prominent leader among women in Utah as well as a prolific writer, a mother of thirteen children, and an ardent supporter of women's suffrage. She founded and edited both the Young Woman's Journal *and the* Relief Society Magazine *and authored the chapter on Utah women's suffrage work in the final volume of NAWSA's* History of Woman Suffrage. *Gates served as an officer of the National Council of Women and spoke at the 1893 Chicago World's Fair and at several conferences of the International Council of Women. She helped organize the National Council of Women Voters and was honored at the NAWSA Jubilee Convention in 1920.*

1911

The National Council of Women Voters was organized on January 14 in Tacoma, Washington, with Susa Young Gates as the representative for Utah. This nonpartisan coalition of the five equal-suffrage states (Wyoming, Colorado, Utah, Idaho, and Washington) was organized to educate voting women and to promote suffrage in further states.

The UCW formed a women's Legislative Council with representatives from key organizations. They met during each Utah legislative session to review every bill relevant to women's rights and interests. The organization still exists today.

California became the sixth state to adopt women's suffrage when voters narrowly approved a state constitutional amendment in a special election held October 10. California suffragists cited positive results in Utah and the other enfranchised states to gain support for suffrage.

On November 7, Tamar Hamblin, Luella McAllister, Blanche Hamblin, and Vinnie Jepson were elected as members of the Kanab Town Council with Mary Chamberlain elected as council chair (mayor). Although initially put on the ballot as a joke, they accepted their election and served for the 1912–1913 term, earning them a place in history as the first all-woman town council in Utah.

THE STEAM ROLLER

1912

Annie Wells Cannon, Dr. Jane Skolfield, and Annie H. King were elected to the Utah House of Representatives on November 5. Their legislative successes included a mother's pension bill, education bills, a female labor deputy to protect women and child workers, and the nation's first minimum wage for women and minors.

Both Arizona and Kansas gained women's suffrage in elections on November 5. Oregon voters also approved women's voting rights, after previously placing the suffrage question on the ballot five times—more than any other state.

Utah suffragists were "among the most active" participants in the NAWSA convention held in Philadelphia on November 21. Florence Allen of Utah rode in one of the leading limousines to the convention at Independence Hall. Allen had attended speeches by Susan B. Anthony and Anna Howard Shaw in Salt Lake City in 1895 as a child. She became the first woman in the nation to serve on a state supreme court (in Ohio) and later served as Chief Judge of the U.S. Sixth Circuit Court of Appeals.

Margaret Zane Witcher (Cherdron), future leader of the Utah National Woman's Party, represented Utah as a presidential elector in 1912. Witcher was the first woman in the nation to serve in the Electoral College, and she personally met with President William Howard Taft to deliver Utah's votes.

1914 1915 1916

The NAWSA Woman Suffrage Procession was the largest suffrage event in United States history and a turning point in the movement. Organized by Alice Paul to precede the inauguration of President Woodrow Wilson in Washington, D.C., the event attracted large crowds. It included nine bands, four mounted brigades, twenty floats, and an allegorical performance near the Treasury Building. Over two hundred female marchers required emergency care after they were heckled, tripped, and assaulted. Publicity from the parade shifted public opinion, drew attention to the cause, and reinvigorated the suffrage movement.

1913

Suffragists representing the enfranchised states began a "suffrage hike" on February 12, walking 230 miles in seventeen days from New York to Washington, D.C., to join a NAWSA parade.

At least 8,000 NAWSA suffragists marched down Pennsylvania Avenue in Washington, D.C., on March 3 in the first suffrage parade in the nation's capital.

Edna Groshell, a UCW leader who had also served as president of the Woman's Democratic Club of Salt Lake, marched at the head of the Utah contingent in the parade.

1911 1912 1913

Women gained the right to vote in Alaska Territory on March 21, when the territorial legislature unanimously approved women's right to vote in its first act of the 1913 legislative session.

A procession of hundreds of NAWSA suffragists, led by Alice Paul, delivered a petition to Congress on July 31. Utah Senator Reed Smoot spoke in favor of suffrage and presented Utah's portion of the petitions to the Senate.

Eighty-six-year-old Emmeline B. Wells published the last edition of the *Woman's Exponent* in February, ending almost forty-two years of the paper's advocacy for suffrage. It was replaced in 1915 by the *Relief Society Magazine*, an official Relief Society publication edited by suffragist Susa Young Gates.

1914

Utah Senator George Sutherland spoke in support of women's suffrage in the U.S. Senate on February 18.

"SENATOR SUTHERLAND DECLARED THE MOST CONVINCING ARGUMENT FOR WOMAN SUFFRAGE WAS THE LACK OF ANY PERSUASIVE ARGUMENT AGAINST IT."

—OMAHA DAILY BEE, *FEBRUARY 19, 1914*

1914 | 1915 | 1916

Alice Paul (1885–1977), an American student studying in England, adopted militant protest tactics such as picketing and hunger strikes during her participation in the radical wing of the British suffrage movement. After joining NAWSA in 1912 and organizing the 1913 parade, she and Lucy Burns cofounded the Congressional Union for Woman Suffrage (CU) as a NAWSA auxiliary organization to lobby Congress for a federal suffrage amendment. Paul was encouraged by the publicity NAWSA's 1913 parade attracted in Washington, D.C., but NAWSA president Carrie Chapman Catt worried that more extreme tactics would alienate the public. After separating from NAWSA, Paul and Burns went on to lead the National Woman's Party in protests and militant activism. Following passage of the Nineteenth Amendment, Paul drafted the Equal Rights Amendment in 1921 to eliminate discrimination based on gender.

"I AM NOT A SENATOR OF UTAH ALONE, BUT A SENATOR OF THE UNITED STATES AND JUST AS MUCH INTERESTED IN THE WOMEN OF OTHER STATES AS OF UTAH. . . . I SHALL SEE THAT THE WOMAN'S SUFFRAGE BILL IS NEVER MADE RIDICULOUS OR TABLED THROUGH THE CLEVERNESS OF OPPOSING SENATORS."

—SENATOR REED SMOOT

1914

After months of rising tensions and failed negotiations, the Congressional Union officially broke ties with NAWSA in February.

Elizabeth A. Hayward and Lily C. Wolstenholme were elected to the Utah House of Representatives in November. Hayward improved child labor and equal pay for equal work laws and was reelected in 1916.

Suffrage momentum continued to increase as Nevada and Montana joined the ranks of the suffrage states in elections on November 3.

1915

CU leaders arrived in Salt Lake City on August 19. Emmeline B. Wells led an automobile procession of national and local suffrage leaders such as Emily S. Richards, Ruth May Fox, Margaret Zane Cherdron, and Elizabeth Hayward to meet with Senator Reed Smoot to secure his support for a constitutional suffrage amendment.

1914 1915 1916

"ON THE COMMITTEE, AS YOU SEE, ARE GENTILES AND MORMONS; AND REPUBLICANS, A DEMOCRAT, A PROGRESSIVE, AND A NON PARTISAN—SO WE ARE HOPEFUL THAT THE COMMITTEE WILL SUIT ALL FACTIONS."

—ALICE PAUL

A large Utah suffrage convention on August 20 featured speeches from CU organizer Mabel Vernon, convention chairwoman Annie Wells Cannon, and Senator George Sutherland. The convention unanimously adopted Susa Young Gates's motion to pledge support for the CU's work for a federal suffrage amendment.

"WE ARE ALL AMERICANS AND WE WOMEN OF THE EASTERN STATES ARE ENTITLED TO THE SAME RIGHTS AND RECOGNITION AS THE WOMEN OF THE WEST."

—MABEL VERNON

CU leader Alice Paul arrived in Salt Lake City on August 20 due to travel delays. She found much "enthusiasm and interest" and organized a Utah branch of the CU, with Margaret Cherdron as state chair, as well as a state committee including former legislators Annie Wells Cannon, Alice Merrill Horne, and Lily C. Wolstenholme.

The CU held a convention of voting women in San Francisco on September 14. Annie Wells Cannon attended as vice-chairwoman, along with several other Utah delegates. The convention produced an 18,333-foot-long petition, including 500,000 signatures of western women from the suffrage states. Sara Bard Field from California was selected to lead an envoy to transport the petition across the country and present it to Congress and to President Woodrow Wilson.

1917 | 1918 | 1919

1915

The CU suffrage envoy arrived in Salt Lake City on October 4. They were greeted by the governor, the Salt Lake City mayor, Emmeline B. Wells, and many other public officials and Utah suffragists in an elaborate parade and official ceremony on the steps of the new Utah State Capitol building.

The CU suffrage envoy arrived in Washington, D.C., on December 6, accompanied by a dramatic parade. Lily C. Wolstenholme from Utah was among the suffragists who met with President Wilson and delivered the immense suffrage petition.

Senator George Sutherland from Utah welcomed the suffragists on the steps of the U.S. Capitol and then introduced the "Susan B. Anthony Amendment" to the Senate. Sutherland advised Alice Paul on the wording for the Equal Rights Amendment in 1921 and was later appointed to the U.S. Supreme Court in 1922.

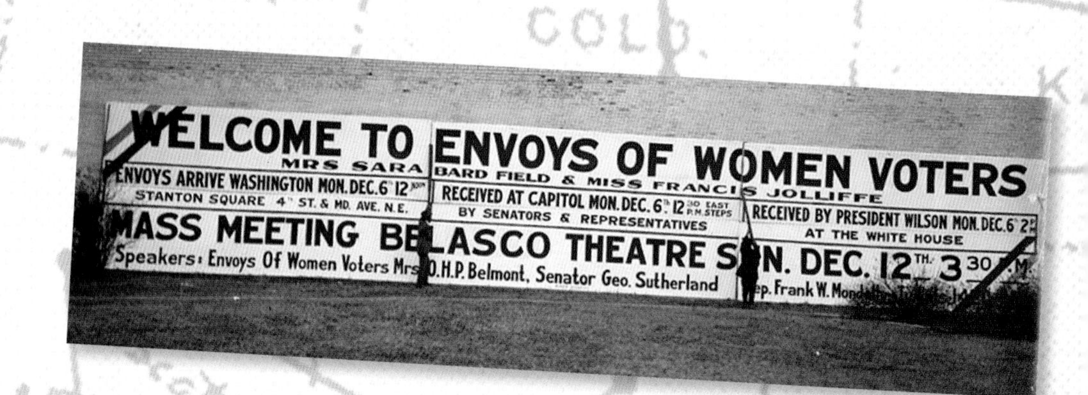

1913 1914 1915

1915

The CU held its first national convention in Washington, D.C., in December. Margaret Zane Cherdron, state chairwoman for Utah, presided over a large convention meeting on December 13. Utah suffragists Lily C. Wolstenholme and Jeanette Young Easton also attended.

1916

Twenty-three CU leaders left Washington, D.C., on April 9 for a five-week train tour to garner support among women voters for a federal woman suffrage amendment. They completed their transcontinental suffrage tour with a convention in Salt Lake City held May 11–12.

Utah suffragist Alice Reynolds joined the CU's "Suffrage Special" envoy of voting women from the west who traveled back to Washington, D.C. The envoy presented resolutions and a large petition for a federal suffrage amendment to an assembly of senators and congressmen in the Capitol Rotunda on May 16.

1916 1917 1918

1916

A Utah branch of the National Council of Women Voters was formed on May 22 with Lily C. Wolstenholme as president. There were soon local branches in Ogden, Brigham City, Logan, and Provo.

The CU held a convention for enfranchised women in Chicago on June 5 and formed the National Woman's Party (NWP), briefly known as the Woman's Party of Western Voters. The NWP replaced the CU in suffrage states such as Utah, while the CU continued in other states.

Annie Wells Cannon, Lily C. Wolstenholme, and Margaret Zane Cherdron represented Utah at the first NWP convention August 10–11. Beginning in September, the NWP held weekly meetings at its Utah state headquarters.

NAWSA President Carrie Chapman Catt unveiled her "Winning Plan" on September 7. Her strategy called for the more than two million NAWSA members to simultaneously work toward a federal suffrage amendment and a state-by-state approach to expand suffrage support.

1913 1914 1915

So anemic that she could barely stand, CU leader Inez Milholland greeted a receiving line of over one hundred women in Ogden on October 17 and then spoke to an overfilled theater in Salt Lake City that night, declaring "women will stand by women." These were some of her last speeches before she collapsed on stage and died in California a few weeks later at the age of thirty.

Zitkála-Šá, a leading advocate for Native people's citizenship and civil rights, moved from Utah to Washington, D.C., in 1916. There, she gave public lectures on Native Americans' cultural identity and promoted a pan-Indian movement to work for Native citizenship rights.

On November 7, Elizabeth Hayward, Dr. Grace Stratton-Airey, and Daisy C. Allen were elected to the Utah House of Representatives. Stratton-Airey served two terms, both as the chair of the Public Health Committee. She had served as a delegate to the NAWSA convention in Colorado in 1916 and as a delegate to the Utah state Democratic convention.

Zitkála-Šá (1876–1938), an accomplished pianist, violinist, and lecturer, lived on the Uintah-Ouray Reservation in Utah for fourteen years. She joined the Society of American Indians, working both to preserve the Native way of life and to lobby for Native people's right to full American citizenship, and then continued her activism in Washington, D.C., beginning in 1916. Her efforts, and those of many others working for Native people's rights, bore fruit in 1924 when Native Americans gained U.S. citizenship through the Indian Citizenship Act. Still, many states, including Utah, made laws and policies that prohibited Native Americans from voting for decades to come. In 1926, Zitkála-Šá cofounded the National Council for American Indians to continue to lobby for Native suffrage rights. She served as the council's president, public speaker, and major fundraiser until her death in 1938. It was not until February 14, 1957, that the Utah state legislature finally repealed the law that had prohibited Native Americans living on reservations from voting.

Known as the "Silent Sentinels of Liberty," almost two thousand women from the NWP participated in picketing in front of the White House, six days a week, for over two years. The protest lasted from January 10, 1917, until the U.S. Congress passed the women's suffrage amendment on June 4, 1919. While the protesters were supported by some onlookers, they were harassed, ignored, and attacked by others. Beginning in June 1917, nearly five hundred women were arrested on charges of obstructing traffic and 168 women served jail time. Two Salt Lake City women were among thirty-three suffragists who experienced beatings and torture during the infamous "Night of Terror" in a Northern Virginia prison. Although the picketing protest was considered radical, the mistreatment of the Silent Sentinels angered many Americans and created sympathy for the suffrage movement. By the time the suffrage amendment became law in 1920, a U.S. Court of Appeals had cleared the Silent Sentinels of all charges for their civil disobedience, creating an important legal precedent that paved the way for generations of protestors.

1917

The NWP began an unprecedented and highly controversial picketing campaign at the White House on January 10. The protestors carried large banners that accused President Wilson of hypocrisy for supporting democracy across the world while leaving American women without rights.

The United States entered World War I on April 6. Despite women's previous advocacy for peace, NAWSA and its local branches rallied around the war effort. Approximately eighty Utah women served in the war, mostly as nurses. Women's contributions during the war helped persuade lawmakers that women deserved suffrage.

1912 1913 1914

WOMEN VOTE

IN WYOMING, IDAHO, COLORADO, UTAH, WASHINGTON AND CALIFORNIA

WHY NOT IN NEW YORK?

Maud Fitch of Eureka, Utah, bought her own ambulance and fuel, shipped it overseas, and volunteered as a battlefield ambulance driver in France during the war. Fitch was awarded the French Cross for her service.

Utah suffragist Lily C. Wolstenholme served as president of the National League of Women's Services during WWI and led hundreds of women supporting war work. She organized a parade in downtown Salt Lake City to help with army recruiting.

Emily S. Richards, as President of the UCW, was among the one hundred suffragists to join Carrie Chapman Catt at the White House to meet with President Wilson on October 26 and offer the wartime support of over two million NAWSA members.

New York became the first eastern state to enfranchise women on November 6.

1915 1916 1917

1917

Salt Lake City suffragists Minnie Quay and Lovern Robertson joined the NWP "Silent Sentinel" protest on November 10 and were among the dozens of suffragists arrested while picketing the White House that day.

UCW president Emily S. Richards and Democratic Woman's Club president Hortense Haight Nebeker, both from NAWSA affiliate organizations, publicly denounced the NWP protest as too radical and ineffective. They reaffirmed their loyalty to President Wilson, as well as to the NAWSA policy opposing "the tactics of this militant group of women." In a published letter, they condemned Minnie Quay's participation and terminated her membership in the Woman's Democratic Club. They did not mention Lovern Robertson, likely because she was a Socialist and not a member of their organizations.

Quay, Robertson, and the other protestors from November 10 were sentenced to prison and held in the Occoquan Workhouse. On November 15, they were victims of the "Night of Terror," during which guards treated the imprisoned protestors with violence.

"I AM READY TO DO ANYTHING WITHIN MY POWER AND NO SACRIFICE IS TOO GREAT."

—*MINNIE QUAY*

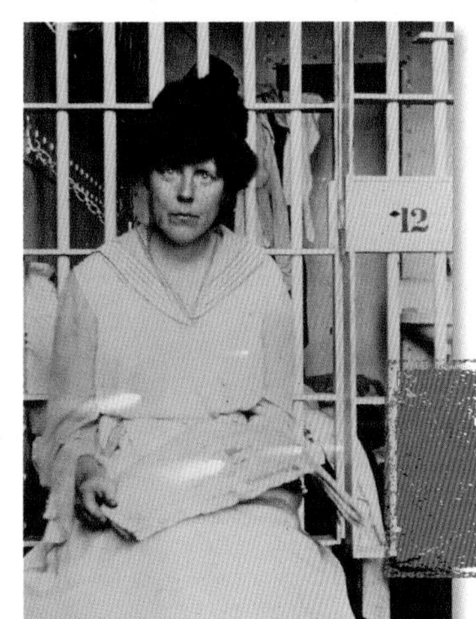

"I DO NOT EXPECT TO ESCAPE ARREST."

—*LOVERN ROBERTSON*

1915 1916 1917

1918

In a dramatic reversal, President Woodrow Wilson urged Congress on September 30 to pass the Nineteenth Amendment, changing his earlier position that states should decide on women's suffrage individually. The amendment did not pass during that Congress.

On November 5, voters in Michigan, South Dakota, and Oklahoma each approved state constitutional amendments granting suffrage to women.

Elizabeth Hayward was elected to the Utah Senate on November 5, and Anna G. Piercey and Delora W. Blakely were elected to the Utah House of Representatives. Dr. Grace Stratton-Airey was reelected as a representative.

"WE HAVE MADE PARTNERS OF THE WOMEN IN THIS WAR. . . . SHALL WE ADMIT THEM ONLY TO A PARTNERSHIP OF SUFFERING AND SACRIFICE AND TOIL AND NOT TO A PARTNERSHIP OF PRIVILEGE AND RIGHT?"
—PRESIDENT WILSON

1919

Reverend George W. Harts, minister of the Calvary Baptist Church, formed the Salt Lake chapter of the NAACP on February 10 in response to racial discrimination in local schools. From the start, African American women played a critical role in the battle for political and civil rights.

NAWSA held its Golden Jubilee Convention in St. Louis, Missouri, March 23–29 to celebrate the organization's fiftieth anniversary. Professor Alice Reynolds from Brigham Young University attended as a delegate representing Utah.

1918 1919 1920

1919

The U.S. Senate passed a constitutional amendment for women's suffrage on June 4, two weeks after it had passed the House of Representatives. The amendment was now up to the state legislatures—once thirty-six states voted to ratify, it would become the Nineteenth Amendment to the U.S. Constitution.

Utah women held a memorial service on July 6 in the Assembly Hall for former NAWSA president Dr. Anna Howard Shaw, who had passed away on July 2. Those at the service unanimously voted to again ask Governor Bamberger to call a special legislative session to ratify the Nineteenth Amendment.

Abby Scott Baker, the NWP political chair, came to Salt Lake City in August during a state governors' conference and urged the governors to call special legislative sessions to ratify the Nineteenth Amendment.

SPECIAL LEGISLATURE FOR SUFFRAGE ASKED

Women at Dr. Shaw Memorial Service Request Action by Governor.

Speakers Laud Dead Leader's Strong Stand for League of Nations.

At the close of the memorial services for Dr. Anna Howard Shaw, held yester-

voice a living tribute to the memory of the former leader of the suffrage cause in whose honor the services were held, as it showed that the women of Utah are now, as they ever have been since suffrage was granted them in 1896, loyal supporters of the suffrage cause.

Glowing tributes of respect to Dr. Shaw w[...] given by several speakers.

"I APPEAR IN THE ADVOCACY OF A GREAT CAUSE, WITH THE TIME RIPE, WITH THE PEOPLE READY, WITH THE BENEFICIARIES EAGER, INTELLIGENT, AND WILLING TO PERFORM THEIR SHARE IN THE FUNCTIONING OF GOOD GOVERNMENT."

—JAMES R. MANN, CHAIRMAN OF THE HOUSE WOMAN SUFFRAGE COMMITTEE

On September 23, President Woodrow Wilson spoke in the Salt Lake Tabernacle on the merits of the League of Nations. While in Utah, Wilson paid only one official visit—to thank Emmeline B. Wells for her leadership in the grain-saving effort and for selling 200,000 bushels of grain to the United States military during World War I.

On September 29, State Senator Elizabeth Hayward introduced a bill to ratify the Nineteenth Amendment into the special session of the Utah Senate. Representative Anna T. Piercey chaired the House during the passage of the amendment on September 30, and Representatives Delora W. Blakely and Dr. Grace Stratton-Airey gave speeches.

Utah officially ratified the Nineteenth Amendment on October 3, becoming the first of the suffrage states to ratify the amendment and the seventeenth state overall.

Carrie Chapman Catt visited Salt Lake City to attend a convention of the Utah League of Women Voters held November 17–18. Catt and other NAWSA leaders gave public addresses at the Assembly Hall, Tabernacle, and Hotel Utah.

1918 1919 1920

1920

Utah women planned a jubilee celebration for February 12, 1920, in honor of the fiftieth anniversary of Utah women gaining the vote, with speeches, music, and an exhibit of Utah suffragists' portraits.

NAWSA held a "victory convention" in Chicago in February to officially dissolve the suffrage association into the League of Women Voters. Susa Young Gates and Donette Smith Kesler of Utah were among the two thousand delegates who celebrated the Nineteenth Amendment's progress. The convention honored suffrage pioneers including Utah's Emmeline B. Wells, Emily S. Richards, and Susa Young Gates.

The Tennessee state legislature narrowly voted to ratify the Nineteenth Amendment on August 18, 1920, the final ratification needed for the amendment to become national law. After two tied votes, in which state representative Harry T. Burn had voted against the amendment, he changed his mind and cast the deciding vote in favor of suffrage because of a timely letter from his mother, Febb Burn.

"A MOTHER'S ADVICE IS ALWAYS SAFEST FOR HER BOY TO FOLLOW, AND MY MOTHER WANTED ME TO VOTE FOR RATIFICATION."

—*HARRY T. BURN*

"HURRAH, AND VOTE FOR SUFFRAGE."

—*FEBB BURN*

1915 1916 1917

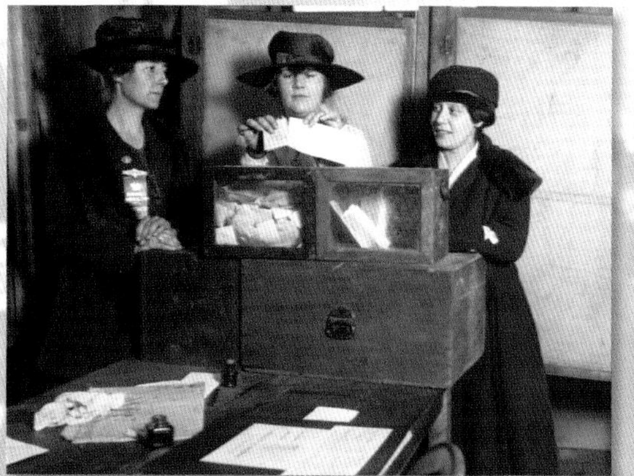

The Nineteenth Amendment officially became part of the U.S. Constitution when Secretary of State Bainbridge Colby signed a proclamation on August 26, 1920. There was no ceremony, but women across America celebrated.

Utah women celebrated with a parade, bands, and a program with speakers on the steps of the State Capitol on August 31.

Millions of women across the country voted for the first time in the general election on November 2. While the Nineteenth Amendment was a major step toward women's political equality, much work remained to provide every American woman a truly equal opportunity to cast a ballot and exercise her political voice.

1918

1919

1920

UTAH SUFFRAGIST RUTH MAY FOX CASTING A BALLOT IN 1956.

EPILOGUE

"LET US NOT WASTE OUR TALENTS IN THE CAULDRON OF MODERN NOTHINGNESS, BUT STRIVE TO BECOME WOMEN OF INTELLECT, AND ENDEAVOR TO DO SOME LITTLE GOOD WHILE WE LIVE IN THIS PROTRACTED GLEAM CALLED LIFE."

—DR. MARTHA HUGHES CANNON

EMMELINE B. WELLS WROTE IN 1911 that she hoped future historians would "remember the women of Zion when compiling the history of this Western land." This book is a piece of that process. Utah's unique experience within the suffrage movement revealed tensions and patterns that illuminate the larger suffrage story, and the women of Utah demonstrated time and again that they were deliberate, invested, and active players within the suffrage movement.

The passage of the Nineteenth Amendment was an important step toward equality for women in America. Gaining the right to vote was the most controversial—and arguably the most important—of the goals articulated by Elizabeth Cady Stanton, Lucretia Mott, and others at that first American women's rights convention in 1848. It had taken more than seventy years of speeches, petitions, conventions, lobbying, parades, and picketing to realize their dream of nationwide women's suffrage. Only one person who had signed Stanton's "Declaration of Sentiments" was still alive in 1920 when the Nineteenth Amendment was ratified, but a second and third generation of women and men had risen to work for the cause in New York and Utah and across the nation. Women could no longer be kept from the ballot box simply because they were women.

This achievement is a credit to the women and men whose commitment to equal rights motivated their efforts for women's political, educational, and professional advancement. Utah women were an important part of this national movement, and in many ways their experience paved the way for women's suffrage to extend across the

United States. Not only were Utah women the first to cast ballots since the convention at Seneca Falls, but as they voted, served in public office, educated each other about civics, organized to administer relief, and testified before Congress, they showed the nation that women could be a powerful force for good in every aspect of their communities.

In celebrating the history of suffrage, it is important to remember the Native Americans, immigrants of Asian descent, and people of color, both women and men, whose struggle for voting rights continued beyond 1920. Although the Nineteenth Amendment prohibited gender-based voting restrictions, discriminatory national citizenship laws as well as restrictions imposed by individual states still prevented many people from casting ballots and having their voices heard in government. Native Americans and many Asian immigrants were not eligible to become citizens in 1870 or 1896 or even 1920, meaning they could not legally vote. Legal barriers enacted in many states effectively made it impossible for African Americans to vote. Despite the inequalities they faced, many women and men in these marginalized communities persevered in the struggle for equal voting rights, leading to legislation such as the Indian Citizenship Act of 1924, the Immigration and Nationality Act of 1952, and the landmark Voting Rights Act of 1965, which significantly widened the franchise. This pursuit continues today as many of these communities still face obstacles to truly equal access to the polls.

This book isn't large enough to include all the women from diverse communities, professions, and backgrounds who have carried forward Utah's legacy of women's advocacy, organization, and service. Facing war, inequity, apathy, or discrimination, they dug in with grit, humor, and persistence to bring about positive change in their neighborhoods and beyond. Just like Utah's nineteenth-century suffragists, they envisioned better days ahead and worked to make those days a reality despite divisions, mistakes, and setbacks. We stand on their shoulders today, and it is incumbent on us to amplify their stories and live up to their legacy.

The year 2020 marks the 150th anniversary of Utah women's first votes and the 100th anniversary of the Nineteenth Amendment. As we honor the achievements of women in the past, we realize that the issues they addressed are still relevant to us today. There is no better time to celebrate women's contributions, take inspiration from Utah's rich cultural heritage of women's advocacy, and resolve to become more engaged participants in our own communities.

IN MEMORIAM

DEBORAH ANN COULAM WHEELWRIGHT
1920–2001

ANN'S LIFELONG LOVE OF books, education, and ideas began when she was four years old, learning to read over her parents' shoulders. For the rest of her life, she was rarely without a book close by. Though she got a jump start on formal education by beginning third grade at age five, Ann worked toward her university diploma for forty-seven years. After some time away, she returned to school in 1944 while her husband, Max, served overseas during World War II, leaving her at home with their two young children. Ann later served for ten years on the Education Committee of the Utah Women's Legislative Council. She kept herself well-informed on local issues and frequently spent the days leading up to elections answering phone calls from women eager to hear her reasoned, respected analysis of the upcoming vote. For Ann, learning was everything: a privilege, a joy, and above all, a responsibility to move beyond books and blackboards to make the world a better place.

ACKNOWLEDGMENTS

We are indebted to the support of colleagues, archivists, friends, family, and volunteers for the creation of this book. First, we express our gratitude to the entire team at Better Days 2020 for their institutional support at every stage. We also offer deep appreciation to Melinda Wheelwright Brown and the Steven C. and Margaret S. Wheelwright Family Foundation for a generous grant that allowed us to make our vision of this book a reality.

Many thanks to helpful staff at The Church of Jesus Christ of Latter-day Saints Church History Library; L. Tom Perry Special Collections, Harold B. Lee Library, Brigham Young University; Marriott Library Special Collections, University of Utah; Utah State Historical Society; and to local and family historians who shared research and images.

Tiffany Greene, Gabi Price, and Emmalyn Pykles provided invaluable research assistance, and this book would not have been possible without their help. Thanks also to Brittany Chapman Nash, Darci Garcia, and Dr. Naomi Watkins for comments that helped shape and improve the content of this book.

We are grateful for the skills of Kate Corkum, Allaina Jeffreys, Alicia Pangman, and Hayley Adams at Re/Mark who did an extraordinary job designing the interior pages of this book so that it is both informational and beautiful. We also offer a heartfelt thank you to Tracy Keck, Celia Barnes, and Shauna Gibby of Deseret Book, who shepherded this unique project through its many twists and turns.

Finally, we thank our husbands, Paul Roberts and Andrew Clark, for their unwavering patience, love, and support.

U.S. WOMEN'S SUFFRAGE TIMELINE

Wyoming Territory	December 10, 1869	(territorial legislature)
Utah Territory*	February 12, 1870	(territorial legislature)
Washington Territory**	November 23, 1883	(territorial legislature)
Wyoming	July 10, 1890	(state constitution)
Colorado	November 7, 1893	(amendment to state constitution)
Utah	January 4, 1896	(state constitution)
Idaho	November 3, 1896	(amendment to state constitution)
Washington	November 8, 1910	(amendment to state constitution)
California	October 10, 1911	(amendment to state constitution)
Arizona	November 5, 1912	(amendment to state constitution)
Kansas	November 5, 1912	(amendment to state constitution)
Oregon	November 5, 1912	(amendment to state constitution)
Alaska Territory***	March 3, 1913	(territorial legislature)
Montana	November 3, 1914	(amendment to state constitution)
Nevada	November 3, 1914	(amendment to state constitution)
New York	November 6, 1917	(amendment to state constitution)
Michigan	November 7, 1918	(amendment to state constitution)
Oklahoma	November 7, 1918	(amendment to state constitution)
South Dakota	November 7, 1918	(amendment to state constitution)

Note: Territorial and state suffrage laws allowed women citizens who met age and residency requirements to vote. Due to discriminatory national citizenship laws, Native American women were not considered U.S. citizens until the Indian Citizenship Act of 1924. After 1924, many states, including Utah, enacted laws prohibiting Native Americans on reservations from voting.

* Utah's suffrage law was repealed by Congress in 1887 through the Edmunds-Tucker Act.

** The Washington Territorial Supreme Court revoked women's suffrage in 1887. A women's suffrage law passed in 1888, but the Territorial Supreme Court ruled later that year that women could not vote.

*** Alaska became a state in 1959.

UTAH WOMEN ELECTED TO THE STATE LEGISLATURE | 1896–1920

Year Elected	Name	Party	City & County	Position	Years Served
1896	Euritha K. LaBarthe	D	Salt Lake City, Salt Lake County	State representative	1 term; 1897–1898
1896	Sarah E. Anderson	D	Ogden, Weber County	State representative	1 term; 1897–1898
1896	Dr. Martha Hughes Cannon	D	Salt Lake City, Salt Lake County	State senator	1 term; 1897–1900
1898	Alice M. Horne	D	Salt Lake City, Salt Lake County	State representative	1 term; 1899–1900
1902	Mary Anna C. Geigus Coulter	R	Ogden, Weber County	State representative	1 term; 1903–1904
1912; 1920	Annie Wells Cannon (Elizabeth Ann Wells Cannon)	R	Salt Lake City, Salt Lake County	State representative	2 terms; 1913–1914; 1921–1922
1912	Ann Holden King	R	Salt Lake City, Salt Lake County	State representative	1 term; 1913–1914
1912	Edyth Ellerbeck Read	R	Salt Lake City, Salt Lake County	State representative	Died on January 20, 1913, before being sworn in
1912	Dr. Jane Manning Skolfield	R	Salt Lake City, Salt Lake County	State representative	1 term; 1913–1914
1914	Elizabeth Pugsley Hayward	D	Salt Lake City, Salt Lake County	State representative	2 terms; 1915–1918
1914	Lily Clayton Wolstenholme	R	Salt Lake City, Salt Lake County	State representative	1 term; 1915–1916
1916	Dr. Grace Stratton Airey	D	Salt Lake City, Salt Lake County	State representative	2 terms; 1917–1920
1916	Daisy C. Allen	D	Salt Lake City, Salt Lake County	State representative	1 term; 1917–1918
1918; 1928	Anna Thomas Piercey	D	Salt Lake City, Salt Lake County	State representative	2 terms; 1919–1920; 1929–1930
1918	Elizabeth Pugsley Hayward	D	Salt Lake City, Salt Lake County	State senator	1 term; 1919–1922
1918	Delora Wilkins Blakely	D	Salt Lake City, Salt Lake County	State representative	1 term; 1919–1920
1920	Cloa Pearl Huffaker Clegg	R	Midway, Wasatch County	State representative	1 term; 1921–1922
1920	May Belle Thurman Davis	R	Salt Lake City, Salt Lake County	State representative	1 term; 1921–1922
1920	Antoinette B. Kinney	R	Salt Lake City, Salt Lake County	State senator	1 term; 1921–1924

UTAH WOMEN ELECTED TO COUNTY OFFICE | 1896-1920

Year Elected	Name	County	Position
1896	Margaret A. Caine	Salt Lake	Auditor
	Ellen Jakeman	Utah	Treasurer
	Delilah K. Olson	Millard	Recorder
	Lottie Farmer	Juab	Recorder
	Maud Layton	Sevier	Recorder
	Mamie Wooley	Kane	Clerk
	Amelia Graehl	Box Elder	Recorder
	Emily Dods	Tooele	Recorder
	Tryphenia West	Iron	Recorder
	Bessie Morehead	Cache	Recorder
	Mary F. Shelby	Rich	Recorder
1898	Eliza Madsen	Box Elder	Recorder
	Mamie Foy	Garfield	Recorder
	Sarah E. Cole	Juab	Recorder
	Perata Spencer	Kane	Recorder
	Annie Wixum	Piute	Recorder
	Kate R. Snowball	Rich	Recorder
	Kate Perkins	San Juan	Clerk/Recorder
	Eliza Ross	Sevier	Recorder
	Mary Herron	Tooele	Treasurer
	May Brown Spencer	Utah	Treasurer
1900	Ella Hutchins	Beaver	Recorder
	Jennie P. Slater	Cache	Recorder
	Sarah A. Howard	Davis	Recorder
	Emily C. Watson	Iron	Recorder
	Hattie Spencer	Kane	Recorder
	Josephine King	Piute	Recorder
	Alveretta Olson	Sevier	Recorder
	Sadie Holdaway	Uintah	Recorder
	Sarah Jane Rogerson	San Juan	Recorder
1902	Rebecca Eames	Cache	Treasurer/Recorder/Clerk
	Mary Worthen	Garfield	Recorder
	Mary Monohan	Beaver	Recorder
	Kathinka Anderson	Emery	Recorder
	Emily C. Watson	Iron	Recorder
	Isabelle Robison	Millard	Recorder
1904	Meltrude Hunsaker	Box Elder	Recorder
	Lydia Cowley	Sevier	Recorder
	Sadie Foss	Davis	Recorder
	Elizabeth "Lizzie" Nielson	Beaver	Recorder
	Christina Madsen	Box Elder	Recorder
	Esta Sorenson	Millard	Recorder
	Laura Stark	Piute	Recorder
	Dorothy Wilson	Rich	Recorder
	Addie Longhurst	Uintah	Recorder
1906	Jennie Ashby	Millard	Recorder
	Minerva Johnson	Uintah	Recorder
	Lucinda Redd	San Juan	Recorder
	Lenore Butt	San Juan	Superintendent of Schools
	Rachel C. Farley	Morgan	Superintendent of Schools
	Sarah Merrill	Cache	Recorder
	Ida M. Wells	Grand	Superintendent of Schools
	Emily Bertelson	Piute	Superintendent of Schools
	Susa J. Searle	Piute	Recorder
	Elizabeth Wilson	Beaver	Recorder
1908	Margaret Zane Witcher	Salt Lake	Clerk/Recorder
	Mary A. Gunn	Iron	Recorder
	Lavina Murdock	Wasatch	Treasurer
	Linnie Telford	Cache	Recorder
	Isabella Dalton	Box Elder	Recorder
	Elsie Somerville	Grand	Clerk
	Hattie Epson	Piute	Recorder
	Pearl Brough	Rich	Recorder
	Viola Burr	Sevier	Recorder
	Ruth Bailey	San Juan	Treasurer
	Lillian Rowberry	Tooele	Superintendent of Schools
1910	Hulda Miller	Davis	Recorder
	Agnes Edwards	Beaver	Recorder
	Rose H. Neeley	Box Elder	Recorder
	Lavina Murdock	Wasatch	Recorder
	May Baker	Sevier	Recorder
	Electa Dorrity	Beaver	Recorder
	Mabel Hacking	Uintah	Recorder
	Kate Preston	Cache	Recorder
	Josie Fitzgerald	Carbon	Recorder
	Ann Cooper	Garfield	Recorder
	Blanche Lewis	Davis	Recorder
	Mary A. Gunn	Iron	Recorder
	Carrie A. Henry	Piute	Recorder
	Emma Christensen	San Juan	Recorder
	Matilda Dalton	Sevier	Recorder
	Ann Snow	Wayne	Superintendent of Schools
1914	Lulu Hood	Duchesne	Treasurer
	Olga Standing	Box Elder	Recorder
	Gwendolyn M. Benson	Iron	Recorder
	Caroline F. Roundy	Kane	Superintendent of Schools
	Alice Davis	Millard	Recorder
	Athena Porter	Morgan	Clerk
	Addie Gottfredson	Piute	Recorder
	Augustine Pearce	Rich	Recorder
1916	Ella Johnson	Box Elder	Recorder
	Clara R. Grey	Sevier	Recorder
	Josie Barnson Sprague	Piute	Recorder
	Francis Callahan	Wayne	Treasurer
	Katherine Higginbotham	Weber	Recorder
	Sophia Mallet	Tooele	Recorder
	Barbara Forrester	Carbon	Recorder
	Louisa Barber	Cache	Recorder
	Jennie Ashbey	Millard	Recorder
	Alma Kimball	Summit	Recorder
	Millie Witt	Wasatch	Treasurer
	Alice M. Todd	Duchesne	Recorder
	Kate Taylor	Iron	Recorder
	Kate Littlefield	Morgan	Clerk/Recorder
	Sarah A. McKinnon	Rich	Recorder
	Zilpha Wall Meeks	Uintah	Recorder
1918	Melissa Lance	Duchesne	Recorder
	Alice Eliason	Box Elder	Recorder
	Mary Ann Thomas	Carbon	Recorder
	Maggie E. Foy	Garfield	Recorder
	Addie L. Swapp	Kane	Recorder
	Bertha Warner	Millard	Recorder
	Mary J. Bowman	Sevier	Recorder
	Lucinda "Lou" Goldbranson	Sevier	Treasurer
	Kate Kimball	Summit	Recorder
	Genevieve Richardson	Utah	Recorder
	Ellen Carter	Washington	Recorder
	Zella Colvin	Wayne	Clerk/Auditor/Recorder
1920	Kate Littlefield	Morgan	Clerk/Recorder
	Ida Parks	Juab	Treasurer
	Ruth Bailey	San Juan	Treasurer
	Isabelle de la Marr	Tooele	Treasurer
	Jenny Jepsen	Utah	Treasurer
	Celista C. Jensen	Box Elder	Recorder
	Laura R. Merrill	Cache	Recorder
	Tessie Sanford	Carbon	Recorder
	Hulda Brown	Davis	Recorder
	Annie M. Allen	Emery	Recorder
	Florence Johnson	Piute	Recorder
	Minnie Cutler	Salt Lake	Recorder
	Blanche Eliason	Sanpete	Recorder
	Minnie Preece	Uintah	Recorder
	Leah Elkins	Utah	Recorder
	Ellen Carter	Washington	Recorder
	Glenn Felt	Weber	Recorder
	Lulu Colvin	Wayne	Recorder
	Elsie Norris	Rich	Recorder
	Mayme Dunchee	Beaver	Recorder
	Gladys Briggs	Wasatch	Treasurer
	Tyrza Hansen	Wayne	Treasurer

SOURCES FOR QUOTATIONS

Epigraph

Emmeline B. Wells [writing as Blanche Beechwood], "Why, Ah! Why," *Woman's Exponent* 3, no. 9 (Sept. 30, 1874): 67.

Introduction

Dr. Martha Hughes Cannon, "Woman Suffrage in Utah," Feb. 15, 1898, U.S. House of Representatives Committee on the Judiciary Hearing on House Joint Resolution 68, Church History Library, The Church of Jesus Christ of Latter-day Saints, Salt Lake City, Utah (hereafter CHL).

CHAPTER 1

Pages 6–7

"Minor Topics," *New York Times*, Dec. 17, 1867, 4.

"The Female Suffrage Question," *Deseret Evening News*, Jan. 9, 1868, 2.

Eliza R. Snow, "Sketch of My Life," Apr. 13, 1885, 1.

Brigham Young, "Remarks," *Deseret News*, Dec. 14, 1867, 2.

Pages 8–9

Elizabeth Cady Stanton to Gerrit Smith, Jan. 1, 1866, Gerrit Smith Collection, Syracuse University Special Collections.

"Female Suffrage in Utah," *Deseret News*, Mar. 24, 1869, 6.

"The Women of Utah," *New York Times*, Mar. 5, 1869, 6.

Pages 10–11

American Equal Rights Association, Henry M. Parkhurst, and Carrie Chapman Catt, *Proceedings of the First Anniversary of the American Equal Rights Association*, National American Woman Suffrage Association Collection (New York: Robert J. Johnston, 1867), 5.

"Woman and Her Mission," *Deseret News*, May 26, 1869, 6.

Pages 12–13

Bathsheba W. Smith and Eliza R. Snow, "Minutes of a Ladies Mass Meeting," Jan. 6, 1870, Fifteenth Ward Relief Society Minutes and Records, CHL.

Pages 14–15

"Great Indignation Meeting of the Ladies of Salt Lake City," *Deseret Evening News*, Jan. 14, 1870, 2; and Jan. 15, 1870, 2.

"The Mormon Women in Council," *New York Herald*, Jan. 23, 1870, 6.

Pages 16–17

Susan B. Anthony, "Petticoat Parliament," *Daily National Republican*, Jan. 19, 1870, 4.

"Mormon Suffrage," *Deseret Evening News*, Feb. 10, 1870, 3.

"Female Suffrage in Utah," *Deseret Evening News*, Feb. 8, 1870, 2.

Pages 18–19

Stephen A. Mann to Orson Pratt, Feb. 12, 1870, in "The Woman Suffrage Bill," *Deseret News*, Feb. 16, 1870, 6.

Pages 20–21

"Ladies' Cooperative Retrenchment Meeting, Minutes," Feb. 19, 1870, Fifteenth Ward Relief Society Minutes and Records, CHL.

S. A. Mann to Eliza R. Snow, Feb. 19, 1870, Brigham Young Office Files, CHL.

"The Ladies' Mass Meetings—Their True Significance," *Deseret News*, Mar. 9, 1870, 1.

Pages 22–23

"The Election," *Salt Lake Herald-Republican*, Aug. 2, 1870, 3.

George A. Smith to Sophronia Lyman, Aug. 3, 1870, Historian's Office Letterpress Copybooks, 1854–1879, 1885–1886, CHL.

Charlotte Cobb Godbe, *Woman's Journal* 7, no. 44 (Oct. 28, 1876): 352.

"William H. Hooper, the Utah Delegate and Female Suffrage Advocate," *Phrenological Journal* 51, no 5 (Nov. 1870): 328.

Paulina W. Davis, *A History of the National Woman's Rights Movement for Twenty Years* (New York: Journeymen Printers' Co-operative Association, 1871), 25.

CHAPTER 2

Pages 26–27

Elizabeth Cady Stanton, in "Woman Suffrage," *Salt Lake Tribune*, July 1, 1871, 3.

Pages 28–29

Eliza R. Snow, "Latter Day Saint Ladies of Utah," *Deseret News*, July 26, 1871, 7.

"Primary Election Meeting," *Salt Lake Herald Republican*, Feb. 4, 1872, 2.

"Illegal Voting," *Salt Lake Tribune*, Feb. 13, 1872, 2.

Pages 30–31

"A Utah Ladies' Journal," *Woman's Exponent* 1, no. 1 (June 1, 1872): 8.

Susan B. Anthony, "Is it a Crime for a Citizen of the United States to Vote?" *Account of the Proceedings on the Trial of Susan B. Anthony* (Rochester, NY: Daily Democrat and Chronicle Book Print, 1874).

Pages 32–33

Eliza R. Snow, in "R. S. Reports," *Woman's Exponent* 4, no. 3 (June 1, 1875): 2.

Wells [writing as Blanche Beechwood], "Why, Ah! Why," 67.

"The Women's Petition," *Deseret News*, Feb. 9, 1876, 1.

"Notes and News," *Woman's Exponent* 4, no. 2 (June 15, 1875): 9.

Pages 34–35

"Woman Suffrage in Utah," *Woman's Exponent* 2, no. 14 (Feb. 15, 1876): 139; and"The Suffrage Movement," *National Republican*, Jan. 29, 1876, 1.

Charlotte Godbe, "Life among the Mormons," *Woman's Journal* 2, no 44 (Oct. 28, 1876): 8.

"Women and the World's Fair," *Woman's Exponent* 21, no. 11 (Dec. 1, 1892): 84.

Pages 36–37

NWSA Petition for Woman Suffrage, in "Woman Suffrage," *Salt Lake Herald-Republican*, Dec. 19, 1877, 3.

Emmeline B. Wells, "Convention Letters," *National Citizen and Ballot Box* 2, July 1877, 1.

Memorial of the Women of Utah, National Archives, "Petitions," Congressional Record, 45th Congress, session 2 (Mar. 4, 1878), 1466.

Lillie Devereaux Blake, "The Proposition to Disfranchise the Women of Utah," *Woman's Exponent* 7, no. 2 (June 15, 1878): 15.

Pages 38–39

"Home Affairs," *Woman's Exponent* 7, no. 6 (Aug. 15, 1878): 44.

Sara Andrews Spencer, in "Our Relief Societies," *Woman's Exponent* 7, no. 10 (Oct. 15, 1878): 76.

"Woman's Mass Meeting," *Woman's Exponent* 7, no. 13 (Dec. 1, 1878): 103.

Margaret A. Cluff, "Mass Meeting," *Woman's Exponent* 7, no. 14 (Dec. 15, 1878): 109.

"Woman's Mass Meeting," 7:13:97.

Pages 40–41

Amanda E. Dickinson, "Polygamy Degrades Womanhood," *Woman's Journal* 10, no. 13 (Mar. 29, 1879): 1.

Pages 42–43

Elizabeth Cady Stanton, "The Brand of the Slave," *National Citizen and Ballot Box*, May 1879.

"Polygamy and Woman Suffrage," *Anti-Polygamy Standard* (June 1880): 20.

"Female Franchise," *Salt Lake Herald-Republican*, Oct. 2, 1880, 3.

Emmeline B. Wells, Diary, vol. 6, July 7, 1881, 121, Brigham Young University (hereafter BYU).

CHAPTER 3

Pages 46–47

Dr. Romania B. Pratt, "Woman's Suffrage Convention," *Woman's Exponent* 10, no. 9 (Mar. 1, 1882): 146.

"Open Letter to the Suffragists of the United States," *Anti-Polygamy Standard* 2, no. 12 (Mar. 1882): 2.

Franklin S. Richards, in Utah State Archives and Records Service, Statehood Constitutional Convention (1895) Records, Series 3212, Mar. 28, 1895.

Pages 48–49

Emmeline B. Wells, "Sweet Is Liberty," *Woman's Exponent* 11, no. 19 (Mar. 1, 1883): 148.
"Principles of the People's Party," 1882, CHL.
"Yesterday," *Salt Lake Tribune*, Nov. 8, 1882, 2.

Pages 50–51

Belva A. Lockwood, "The Disfranchisement of the Women of Utah," *Ogden Daily Herald*, June 9, 1883, 1.
National Woman Suffrage Association, *Report of the Sixteenth Annual Washington Convention* (Rochester, N.Y.: Charles Mann, 1884), 44, Library of Congress (hereafter LOC).

Pages 52–53

Emmeline B. Wells and Annie Godbe, in NWSA, *Report*, 79–80, LOC.

Pages 54–55

"Memorial to the President of the United States and Members of Congress, by the 'Mormon' Women of Utah," 1886, CHL.
Edmunds-Tucker Act, 49th Congress, session 2, ch. 397, Mar. 3, 1887, (24 Stat. 635).
"Thanks to the Woman Suffragists," *Deseret Evening News*, Jan. 31, 1887, 3.

Pages 56–57

History and Minutes of the National Council of Women of the United States, Organized in Washington, D.C., March 31, 1888 (Boston: E.B. Stillings, 1898), 3, 10.
"Editorial Thoughts," *Woman's Exponent* 16, no. 23 (May 1, 1888): 180.

Pages 58–59

"The Sister Suffragists," *Salt Lake Tribune*, Jan. 8, 1889, 4.
Emily S. Richards, in "Utah's Lady Delegate," *Woman's Exponent* 17, no. 18 (Feb. 15, 1889): 137.

Pages 60–61

Elizabeth Ann Schofield, in "W.S.A. in Juab County," *Woman's Exponent* 18, no. 1 (June 1, 1889): 6.
Julia P. M. Farnsworth, in "Woman Suffrage Association" *Woman's Exponent* 17, no. 19 (Mar. 1, 1889): 150.
Hulda Cordelia Smith, in "Morgan County W.S.A.," *Woman's Exponent* 18, no. 3 (July 1, 1889): 21.

CHAPTER 4

Pages 64–65

Sarah M. Kimball, "Greeting," *Woman's Exponent* 18, no. 18 (Feb. 15, 1890): 139.
Elizabeth Cady Stanton, "Change is the Law of Progress," Feb. 12, 1890, Elizabeth Cady Stanton Papers: Speeches and Writings, 1848–1902, LOC.
Susan B. Anthony and Ida Husted Harper, eds., *The History of Woman Suffrage: 1883–1900*, vol. 4 (Rochester, NY: printed by the author, 1902), 942.

Pages 66–67

"Convention Chit-Chat," *Washington Post,* Feb. 23, 1891, 1.
Emily Woodmansee, "Equal Rights," in Utah Woman Suffrage Songbook, 1890, 4, CHL.
Susan B. Anthony telegram to Sarah M. Kimball, in "Anniversary Celebration of Susan B. Anthony's Birthday," *Woman's Exponent* 20, no. 16 (Mar. 1, 1892): 125.
Anthony and Harper, *History of Woman Suffrage: 1883–1900*, 4:943.

Pages 68–69

Dr. Martha Hughes Cannon, in "Utah Women in Chicago," *Woman's Exponent* 21, no. 24 (June 15, 1893): 179.
"Woman's Great Forum," *Chicago Record*, May 15, 1893, 1.
Emily S. Richards and Electa Bullock telegram, in "Utah W.S.A.," *Woman's Exponent* 22, no. 7 (Oct. 15, 1893): 50.
Alvira Lucy Cox, "Equal Suffrage," *Woman's Exponent* 22, no. 7 (Oct. 15, 1893): 50.

Pages 70–71

"Susan B. Anthony's Letter," *Woman's Exponent* 23, no. 3–4 (Aug. 1 and 15, 1894): 169.
"Rawlins and Victory," *Salt Lake Herald-Republican*, Sept. 16, 1894, 6.

Pages 72–73

"S.L. Co. Convention," *Woman's Exponent* 23, no. 11 (Dec. 1, 1894): 211.
Wells, Diary, vol. 19, Feb. 2, 1895, 62, BYU.
Utah State Archives and Records Service, Series 3212, Mar. 14, 1895.
Emmeline B. Wells, "Woman Suffrage Column," *Woman's Exponent* 23, no. 15–16 (Feb. 1 and 15, 1895): 233.

Pages 74–75

Utah State Archives and Records Service, Series 3212, Mar. 18 and 28, 1895.
"Relief Society Conference," *Woman's Exponent* 23, no. 19 (May 1, 1895): 262.
George Q. Cannon, Diary, Apr. 4, 1895, CHL.

Pages 76–77

Joseph F. Smith, in "Relief Society Conference," *Woman's Exponent* 24, no. 6 (Aug. 15, 1895): 44.
Paulina Lyman, in "Woman Suffrage Column," *Woman's Exponent* 24, no. 3 (July 1, 1895): 24.
Ruth May Fox, Diary, Apr. 5, 1895, CHL.
Ruth May Fox, "Lecture on Suffrage," *Woman's Exponent* 24, no. 6 (Aug. 15, 1895): 41–42.

Pages 78–79

Orson F. Whitney and Franklin S. Richards, in Utah State Archives and Records Service, Series 3212, Mar. 28, Mar. 30, and Apr. 5, 1895.
Wells, Diary, vol. 19, Nov. 7, 1895, 340, BYU.
Susan B. Anthony, in "Equal Suffrage in the Constitution." *Woman's Exponent* 23, no. 19 (May 1, 1895): 260.
Fox, Diary, Mar. 18, 1895, CHL.

Pages 80–81

Mary Isabella Horne and Susan B. Anthony, in "Conference N.A.W.S.A.," *Woman's Exponent* 24, no. 11–12 (Nov. 1 and 15, 1895): 77, 79.
Dr. Anna Howard Shaw, in "Conference N.A.W.S.A.," *Woman's Exponent* 24, no. 7–8 (Sept. 1 and 15, 1895): 55.
Sarah M. Kimball, in "Conference N.A.W.S.A.," *Woman's Exponent* 24, no. 9 (Oct. 1, 1895): 61.
Lucy A. Clark, in "Conference N.A.W.S.A.," *Woman's Exponent* 24, no. 20–21 (Apr. 1, 1896): 53.

Pages 82–83

"The New State," *Woman's Exponent* 24, no. 11–12 (Nov. 1 and 15, 1895): 76.
"Utah W.S.A.," *Woman's Exponent* 24, no. 10 (Oct. 15, 1895): 66.

CHAPTER 5

Pages 86–87

Emily S. Richards, in "Woman Suffrage in Utah," *Deseret Weekly*, Feb. 15, 1896, 2.
Susan B. Anthony, "Twenty-Eighth Annual Convention of the N.A.W.S.A.," *Woman's Exponent* 24, no. 14 (Dec. 15, 1895): 95.
Emmeline B. Wells, in "Where Women Vote In Utah," *Political Equality Series* 4, no. 6, New York: NAWSA, Sept. 1899, CHL.
"Non-Partisan Silver Women," *Salt Lake Tribune,* Oct. 13, 1896, 8.

Pages 88–89

Dr. Martha Hughes Cannon, "Our Woman Senator," *Salt Lake Herald-Republican*, Nov. 11, 1896, 5.
Salt Lake Herald-Republican, Oct. 31, 1896, 4.
Dr. Martha Hughes Cannon, "A Woman's Assembly," *Woman's Exponent* 22, no. 15 (Apr. 1, 1894): 114.
Clara B. Colby and Ida Husted Harper, in Anthony and Harper, *History of Woman Suffrage: 1883–1900*, 4:282, 290, 593.

Pages 90–91

Cannon, "Woman Suffrage in Utah," Hearing on House Joint Resolution 68, 12, CHL.
Alice Merrill Horne, "Home and Ideals," *Woman's Exponent* 29, no. 18–19 (Feb. 15 and Mar. 1, 1901): 81.
"Echoes of the Election," *The Broad Ax*, Nov. 12, 1898, 1.

Pages 92–93

May Wright Sewall to Susa Young Gates, Jan. 10, 1899, Susa Young Gates Papers, General Correspondence, National Council of Women, 30, CHL.
Wells, Diary, vol. 24, Feb. 11, 1899, 78, BYU.
Heber M. Wells, quoted in Harriet Horne Arrington, "Alice Merrill

Horne, Art Promoter and Early Utah Legislator," *Utah Historical Quarterly* 58, no. 3 (Summer 1990): 272.

Pages 94–95

Susan B. Anthony, quoted in Kathleen Barry, *Susan B. Anthony: A Biography of a Singular Feminist* (New York: New York University Press, 1988), 342.

Ladies Democratic Club of Provo, "Provo Ladies Protest," *Salt Lake Tribune*, Aug. 25, 1900, 8.

Susa Young Gates, "International Council of Women," *Young Woman's Journal* 10 (Oct. 1899): 435.

Pages 96–97

Blessing upon Emmeline B. Wells, Nov. 9, 1900, Emmeline B. Wells Collection, MSS 805, box 1, folder 5, BYU.

Pages 98–99

"Utah Suffrage Council," *Deseret Evening News,* Apr. 6, 1904, 4.

"Only Woman in the Legislature Made Chairman of Important Committee," *Salt Lake Herald-Republican,* Jan. 14, 1903, 8.

Elizabeth Taylor, "Addresses Her Colored Sisters," *Deseret Evening News,* July 6, 1904, 5.

Pages 100–101

Lucy A. Clark, in "Mrs. Lucy Clark Talks to Women of Chicago," *Inter-Mountain Republican,* June 18, 1908, 7.

"Bits of Information," *Salt Lake Telegram*, Sept. 23, 1908, 10.

"First Woman to Be Seated in Convention," *Ogden Daily Standard,* June 16, 1908, 1.

Pages 102–3

Carrie Chapman Catt, in "Utah Dolls [Featured] At Suffrage Bazar," *Deseret Evening News,* Dec. 23, 1909, 16.

Ida Husted Harper, ed., *The History of Woman Suffrage: 1900–1920*, vol. 6 (New York: NAWSA, 1922), 673.

CHAPTER 6

Pages 106–7

"Women Declare Independence near Liberty Bell," *Salt Lake Herald-Republican*, Nov. 22, 1912, 1.

Pages 108–9

Harper, *History of Woman Suffrage: 1900–1920*, 6:451–53.

"Bristow Says Toll Repeal in Interest of Overland Roads," *Omaha Daily Bee*, Feb. 19, 1914, 1.

Pages 110–11

"Senator Reed Smoot—'Suffragettes are Ready for Convention,'" *Salt Lake Tribune*, Aug. 19, 1915, 12.

Alice Paul to Miss Lancaster, Aug. 23, 1915, Alice Paul Papers, Series II, Suffrage, Congressional Union, General Correspondence, Schlesinger Library, Radcliffe Institute, Harvard University.

Mabel Vernon, in "Women of Utah Pledge Support," *Salt Lake Herald-Republican*, Aug. 21, 1915, 12.

Pages 114–15

Inez Milholland Boissevain, in "Make a Choice, Women Plead," *Salt Lake Herald-Republican*, Oct. 18, 1916, 6.

Pages 118–19

Minnie Quay, in "1000 Pickets Will Heckle Wilson," *Salt Lake Telegram*, Dec. 20, 1917, 2.

Lovern Robertson, in "Salt Lake Woman to Picket White House," *Salt Lake Tribune*, Nov. 1, 1917, 9.

President Woodrow Wilson, Address to the Senate on the Nineteenth Amendment, Sept. 30, 1918, The American Presidency Project, UC Santa Barbara.

Pages 120–21

James R. Mann, in Congressional Record, Proceedings and Debates vol. 53, part 1, House of Representatives, 66th Congress, session 1, May 21, 1919, 88.

Pages 122–23

Harry T. Burn, in "Bribery Attempt Charged in Tennessee Suffrage; Court Attacks Lobbying," *The Evening Star,* Aug. 20, 1920, 1.

Febb Burn to H. T. Burn, Aug. 1920, Harry T. Burn Papers, Calvin M. McClung Historical Collection, Knox County Public Library.

EPILOGUE

Dr. Martha Hughes Cannon to Barbara Replogle, May 1, 1885, Martha H. Cannon Collection, CHL.

Emmeline B. Wells, "The Fortieth Volume," *Woman's Exponent* 40, no. 1 (July 1, 1911): 4.

FURTHER READING

Madsen, Carol Cornwall, ed. *Battle for the Ballot: Essays on Woman Suffrage in Utah.* Logan: Utah State University Press, 1997.

———. *An Advocate for Women: The Public Life of Emmeline B. Wells, 1870–1920*. Provo, Utah: Brigham Young University Press; Salt Lake City: Deseret Book, 2006.

Ulrich, Laurel Thatcher. *A House Full of Females: Plural Marriage and Women's Rights in Early Mormonism, 1835–1870.* New York: Knopf, 2017.

Van Wagenen, Lola. *Sister-Wives and Suffragists: Polygamy and the Politics of Woman Suffrage, 1870–1896*. Provo, Utah: BYU Studies, 2012.

For a full list of sources,
visit www.utahwomenshistory.org/thinkingwomen

IMAGE CREDITS

OPENING TIMELINE

Collage page 1 (top to bottom, left to right)

"Women Marching in Favor of Polygamy," *Frank Leslie's Weekly*, May 31, 1879, Bettman Collection, Getty Images.

Jane S. Richards, Franklin D. Richards family photographs, Church History Library, The Church of Jesus Christ of Latter-day Saints, Salt Lake City, Utah (hereafter CHL).

"Votes for Women" lapel pennant, Library of Congress (hereafter LOC).

Pennsylvania Suffragists on the Picket Line, 1917, National Woman's Party Records, LOC.

Constitution of the National Woman Suffrage Association, 1873, Petitions and Memorials, U.S. House of Representatives Committee on the Judiciary, National Archives.

Hannah Kaaepa in *Young Woman's Journal,* May 1899, 213, CHL.

Victory Map of 1917, New York State Library.

Black women, Utah State Historical Society (hereafter USHS).

"Equal Rights Banner" newsletter, 1893, Beaver County Woman Suffrage Association Papers, L. Tom Perry Special Collections, Harold B. Lee Library, Brigham Young University, Provo, Utah (hereafter BYU).

"The Steamroller," *Judge*, Mar. 17, 1917, LOC.

Collage page 2

"The Awakening," *Puck*, Feb. 20, 1915, 14–15.

Sarah Staker Certificate of Membership of the Woman Suffrage Association of Sevier County, Utah, 1889, Staker Family Collection, CHL.

Utah Gov. William Spry with Emmeline B. Wells and other women, USHS.

Smithfield Branch Relief Society Minutes and Records, 1868–1906, vol. 1, 1868–1878, CHL.

Emmeline B. Wells, USHS.

Zitkála-Šá, 1898, National Museum of American History, Smithsonian Institution (hereafter NMAH), courtesy of Mina Turner.

Martha Horne Tingey in *Young Woman's Journal,* May 1899, 213, CHL.

"Local Membership" ribbon, 1907, Miller NAWSA Suffrage Scrapbooks (hereafter NAWSA Scrapbooks), LOC.

Kanab Ladies' Town Board, USHS.

Chapter 1

An Act in Relation to Women Suffrage [Women Suffrage Act], 1870, Utah Territory Legislative Assembly Papers, 1851–1872, CHL.

Salt Lake City Fifteenth Ward Relief Society Hall, USHS.

Bathsheba W. Smith, Bathsheba Wilson Bigler Smith Portrait Collection, ca. 1800s–early 1900s, CHL.

David Koch, *Seraph Young Votes*, courtesy of the Capitol Preservation Board, Utah State Capitol Collection.

Charlotte Cobb Kirby, Brigham Young Jr. Photograph Collection, ca. 1865–1905, CHL.

Chapter 2

Emmeline B. Wells, Jan. 14, 1879, Emmeline B. Wells Portrait Collection, CHL.

Jennie Froiseth, courtesy of the *Utah Historical Quarterly*.

Thirteenth Ward Relief Society Presidency, ca. 1872, CHL.

Susan B. Anthony and Elizabeth Cady Stanton, ca. 1870, Division of Rare and Manuscript Collections, Cornell University Library.

The Woman's Exponent, June 1, 1872.

Zina Young and Daughter, Zina Diantha Huntington Young Portrait Collection, CHL.

"Votes for Women" flowers, Emma Louise Lyons memorabilia, Historical Society of Washington, DC.

Chapter 3

Jennie Anderson Froiseth letter, Oct. 22, 1881, BYU.

Martha Hughes Cannon, Utah State Legislators Photograph Collection, USHS.

"Equal Rights Banner," BYU.

Sarah M. Kimball, USHS.

"Meeting Under the Auspices of the National Woman's [Woman] Suffrage Association at Farwell Hall, June 1st," *Frank Leslie's Illustrated Newspaper,* June 19, 1880, 265.

Emily S. Richards, USHS.

Chapter 4

Emma McVicker, *Salt Lake Tribune* files, courtesy of Jeremy Harmon.

Lucy A. Rice Clark, USHS.

National Council of Women delegates, *Young Woman's Journal,* June 1895, 390.

President Woodruff's Manifesto, 1890, CHL.

Constitution of the State of Utah, Article 4, May 8, 1895, Utah State Archives.

Susan B. Anthony with Women Suffrage Leaders, 1895, CHL.

Salt Lake County Woman Suffrage Association Membership Ticket, 1891, Bathsheba W. Smith Collection, CHL.

Chapter 5

Utah State Senate, 1897, USHS.

Martha Hughes Cannon, USHS.

National American Woman Suffrage Association Delegate ribbon, 1902, NAWSA Scrapbooks, LOC.

Salt Lake Tabernacle with Utah star, USHS.

Elizabeth Taylor in *Utah Plain Dealer*, courtesy of Josephine Taylor Dickey and Amy Tanner Thiriot.

Portrait of Mary A. Burnham Freeze, CHL.

Utah postcard, Catherine H. Palczewski Postcard Archive, University of Northern Iowa, Cedar Falls, IA.

Susa Young Gates, USHS.

Ruth May Fox, courtesy of Brittany Chapman Nash.

Chapter 6

Suffragists lobby Senator Smoot, 1915, National Woman's Party Collection, Belmont-Paul Women's Equality National Monument, Washington DC (hereafter NWP).

Elizabeth Pugsley Hayward, courtesy of the Wessman family.

Suffragists celebrate, Special Collections Department, Bryn Mawr College Library (hereafter Bryn Mawr).

Joint Resolution of Congress proposing a constitutional amendment extending the right of suffrage to women, May 19, 1919, National Archives.

National Woman's Party Utah Headquarters, 1916, NWP.

Parade of Utah women voters on Main Street, Salt Lake City, 1915, NWP.

Picket Line of Nov. 10, 1917, NWP Records, LOC.

Collage page 3

YLMIA program for World's Congress of Representative Women, 1893, A. Elmina Shepard Taylor Collection, CHL.

Zina Young Card in *Young Woman's Journal,* May 1899, 213, CHL.

"Conquerors," *Judge,* Nov. 1, 1913.

Prescinda [Presendia] Huntington Kimball, Bathsheba W. Bigler Smith Photograph Collection, CHL.

Seraph Young, USHS.

Women Suffrage Leaders, USHS.

Woman Suffrage Button, ca. 1910, Division of Political and Military History, NMAH.

Woman Suffrage Flag, ca. 1900, NMAH.

Utah Woman Suffrage Songbook, 1890, CHL.

FOREWORD

Courtesy of Christine M. Durham.

INTRODUCTION

Women Voters Envoys, NWP.

CHAPTER 1

David Koch, *Seraph Young Votes*, Utah State Capitol Collection.

Pages 6–7

Background: Deseret News [Office], used by permission, USHS.

"The Female Suffrage Question," *Deseret Evening News*, Jan. 9, 1868, 2.

Eliza R. Snow, 1866, Bathsheba W. Bigler Smith Photograph Collection, CHL.

American Fork Ward Relief Society Minutes and Records, 1868–1901, vol. 1, 1868–1880, CHL.

Smithfield Branch Relief Society Minutes and Records, 1868–1906, CHL.

Leading Women of Zion, ca. 1867, CHL.

Pages 8–9

Background: *Deseret News*, Mar. 24, 1869, 6.

Rep. George Julian, LOC.

William S. Godbe, USHS.
"The First Vote," *Harper's Weekly*, Nov. 16, 1867, LOC.
"Female Suffrage," *Frank Leslie's Illustrated Newspaper*, Oct. 2, 1869, 56.
H.R. 64, 1869, CHL.
William H. Hooper, USHS.

Pages 10–11
Background: *East and West Shaking Hands at Laying Last Rail*, 1869, Beinecke Rare Book and Manuscript Library, Yale University.
"Woman's Rights Convention," *Harper's Weekly*, June 11, 1859.
Suffragists with NWSA banner, LOC.
Background: *Deseret News*, May 26, 1869, 6.
George Q. Cannon, George Q. Cannon Portrait Collection, CHL.
Salt Lake City Fifteenth Ward Relief Society Hall, USHS.
Lucy Stone, LOC.

Pages 12–13
Background: Wyoming Territory map, ca. 1882, Michael Cassity Collection, Wyoming State Historic Preservation Office.
Charlotte Cobb Godbe, USHS.
Augusta Adams Young, CHL.
An Act to Grant to the Women of Wyoming Territory the Right of Suffrage and to Hold Office, LOC.
Background: "Minutes of a Ladies Mass Meeting," Jan. 6, 1870, Fifteenth Ward Relief Society Minutes and Records, CHL.
Bathsheba Smith, Bathsheba W. Bigler Smith Photograph Collection, CHL.
Sarah M. Kimball, Hiram Kimball family, CHL.

Pages 14–15
"Women Marching in Favor of Polygamy," Getty Images.

Pages 16–17
Susan B. Anthony, LOC.
Utah Territorial Legislature—House (1866–67), USHS.
An Act in Relation to Suffrage [Women Suffrage Act], CHL.
Group of Miners and Women, Eureka (Utah) Photograph Collection, USHS.
Orson Pratt, Historic Scenes and Portraits, ca. 1880s–1890s, CHL.
"Female Suffrage in Utah," *Deseret Evening News*, Feb. 8, 1870, 2.

Pages 18–19
Background: An Act in Relation to Women Suffrage [Women Suffrage Act], CHL.
"An Act Conferring Upon Women the Elective Franchise," *Acts, Resolutions and Memorials of the Territory of Utah, Passed at the Nineteenth Annual Session of the Legislature, 1870*, 8, Utah State Constitution and Historical Statutes Collection, J.Willard Marriott Library, University of Utah.
Background: Salt Lake City, City Hall, USHS.
Tenth Ward Band, Charles W. Symons Photograph Collection, CHL.
Seraph Young, USHS.
Five Generations of Voting Mormon Women, ca. 1920, CHL.

Pages 20–21
Sarah M. Kimball, USHS.
Phoebe Carter Woodruff, 1866, Woodruff Portraits, CHL.
Prescinda [Presendia] Huntington Kimball, CHL.
Margaret T. Smoot, Abraham O. Smoot and wives, ca. 1860s–1870s, CHL.
Wilmirth East, courtesy of Janalee McBride.
Bathsheba W. Smith, Bathsheba W. Bigler Smith Photograph Collection, CHL.
S. A. Mann to Eliza R. Snow, February 19, 1870, Brigham Young Office Files, CHL.

Pages 22–23
"The Election," *Salt Lake Herald-Republican*, Aug. 2, 1870, 3.
"Woman Suffrage in Wyoming Territory—Scene at the Polls in Cheyenne," *Frank Leslie's Illustrated Newspaper*, Nov. 24, 1888, LOC.
Fifteenth Ward Relief Society, ca. 1887, CHL.
Mary Ashton Rice Livermore, in *Our Army Nurses*, Mary A. Gardner Holland, 1895, 36.
Paulina W. Davis, LOC.
Charlotte Cobb Kirby, CHL.

CHAPTER 2

Emmeline B. Wells, CHL.

Pages 26–27
Background: George S. Bowen letter, June 10, 1871, Brigham Young Office Files, CHL.
Susan B. Anthony, USHS.
"A Lady Delegate Reading Her Argument in Favor of Woman's Voting," *Frank Leslie's Illustrated Newspaper*, Feb. 4, 1871, 349, LOC.
Susan B. Anthony and Elizabeth Cady Stanton, Cornell University Library.
Background: Godbe House, Photographs of Salt Lake City, ca. 1865, CHL.
Old Tabernacle, Temple Block, Salt Lake City, Daguerreotype Collection ca.1850s–1860s, CHL.
Salt Lake Collegiate Liberal Institute, USHS.
Elizabeth Cady Stanton, in *Eminent Women of the Age*, James Parton et al., 1869, 333.

Pages 28–29
Background: Tremont Temple (interior), LOC.
Eliza R. Snow, Bathsheba W. Bigler Smith Photograph Collection, CHL.
Charlotte Cobb Godbe, USHS.
Background: "Primary Election Meeting," *Salt Lake Herald Republican*, Feb. 2, 1872, 2.
Thirteenth Ward Relief Society Presidency, CHL.
Elizabeth Howard, USHS.

Pages 30–31
Background: "Petition," *Deseret News*, May 22, 1875, 6.
Louisa Lula Greene Richards, Engraved Portrait Collection, ca. 1890, CHL.
The Woman's Exponent, June 1, 1872.
Phoebe Couzins, in *History of Woman Suffrage, vol. 3, 1876–1885*, eds. Elizabeth Cady Stanton, Susan B. Anthony, and Matilda Joslyn Gage, 1887, frontispiece.
"Mrs. Woodhull Asserting Her Right to Vote," The Miriam and Ida D. Wallach Print Collection, New York Public Library (hereafter NYPL).
Background: Petition to Congress from the National Woman Suffrage Association, 1873, Petitions and Memorials, U.S. House of Representatives Committee on the Judiciary, National Archives.
"The Woman Who Dared," *The Daily Graphic*, June 5, 1873, LOC.
Constitution of the National Woman Suffrage Association, National Archives.

Pages 32–33
Background: Salt Lake Theatre, ca. 1865, CHL.
Emmeline B. Wells, Emmeline B. Wells Portrait Collection, CHL.
Victoria Woodhull, Historical Photographs and Special Visual Collections, Harvard Art Museum.
Background: "Memorial to Congress by 26,626 Women of Utah," *Deseret News*, Jan. 26, 1876, 7.
Virginia Minor, LOC.
Mrs. C[aroline]. M Severance, Carrie Chapman Catt Papers, Bryn Mawr.
President Ulysses S. Grant, LOC.

Pages 34–35
National Republican, Jan. 29, 1876, 1.
Belva Lockwood, LOC
"Utah Women Suffrage," *Salt Lake Herald Republican*, Feb. 29, 1876, 2.
Emmeline B. Wells to John Taylor, Nov. 16, 1877, First Presidency Correspondence, 1877–1887, CHL.
Background: Brigham Young to Emmeline B. Wells, Nov. 13, 1876, Brigham Young Office Files, CHL.
Emmeline B. Wells in *Three Women of Mormondom*, ca. 1876, CHL.
The Woman's Exponent, July 1, 1877.

Pages 36–37
Background: United States Capitol, LOC.
Emmeline B. Wells, USHS.
National Woman Suffrage Association, Appeal for a Sixteenth Amendment, Nov. 10, 1876, National Archives.
"Female Suffrage," *Salt Lake Herald Republican*, Dec. 15, 1877, 3.
Background: "The Utah Question Once More," *Women's Exponent*, Jan. 15, 1879, 126.
Hon. George Q. Cannon of Utah, LOC.
Lillie Devereux Blake, LOC.

Pages 38–39
Background: Independence Hall, USHS.
Emmeline B. Wells, Emmeline B. Wells Portrait Collection, CHL.
"An Unsightly Object," *Judge*, Jan. 28, 1882, LOC.
Jennie Froiseth, courtesy of the *Utah Historical Quarterly*.
Utah Stake Relief Society, ca. 1892, CHL.

Pages 40–41
Background: *The Woman's Journal*, Mar. 29, 1879, Schlesinger Library, Radcliffe Institute, Harvard University (hereafter Schlesinger Library).
Zina Young and Daughter, CHL.
Matilda Joslyn Gage, in *History of Woman Suffrage vol.1, 1848–1861*, eds. Elizabeth Cady Stanton, Susan B. Anthony, and Matilda Joslyn Gage, 1881, 753.
Supreme Court of the United States, ca. 1886, LOC.
President and Mrs. [Lucy] Rutherford B. Hayes, LOC.
Address of Elizabeth Cady Stanton, LOC.
Zina Y. Williams, USHS.

Pages 42–43
Background: *Anti-Polygamy Standard*, June 1880, 1, BYU.
Elizabeth Cady Stanton, LOC.
The Woman's Exponent, Nov. 1, 1879, 1.
Background: Salt Lake County Courthouse, USHS.
"To Disenfranchise Women," *Salt Lake Herald Republican*, Sept. 26, 1880, 4.
Joseph F. Smith, Joseph F. Smith Portrait Collection, ca. 1884, CHL.
Zina D. H. Young, Bathsheba W. Bigler Smith Photograph Collection, CHL.

CHAPTER 3

"Meeting Under the Auspices of the National Woman's [Woman] Suffrage Association at Farwell Hall, June 1st," *Frank Leslie's Illustrated Newspaper*, June 19, 1880, 265.

Pages 46–47
Background: *Anti-Polygamy Standard*, Mar. 1882, 1, CHL.
Zina Diantha Huntington Young, Cartes-de-visite Portrait Collection, CHL.
"Open Letter to Suffragists," *Anti-Polygamy Standard*, Mar. 1882, 1, BYU.
"The Great Sin of the Century," *Daily Graphic*, Nov. 21, 1883, BYU.
Anti Polygamy Autograph Quilt, 1882, Church History Museum, Salt Lake City, Utah.
Utah Commission Voter Registration Oath, courtesy of Ron Fox.
Franklin Snyder Richards and Emily S. Tanner Richards, USHS.

Pages 48–49
Great Salt Lake County Courthouse, George A. Smith Photograph Collection, CHL.
Deseret Hospital Board of Directors, ca. 1882, CHL.
Background: Principles of the People's Party, 1882, CHL.
International Council of Women Participants, Newseum, Culver Pictures.
"Women Voting in the Municipal Election in Boston," *Harper's Weekly*, Dec. 15, 1888, 965.

Pages 50–51
Background: United States Capitol, in *Anecdotes of Abraham Lincoln and Lincoln's Stories*, Abraham Lincoln, J.B. McClure, ed., 1884, 84.
Belva Lockwood, LOC.
John A. Logan of Illinois, LOC.
Alice Stone Blackwell, LOC.
Background and image: "The Mormon Question," *Daily Graphic*, Oct. 22, 1883, BYU.
Angie F. Newman, in *Collection of Nebraska Pioneer Reminiscences*, Nebraska Society of the Daughters of the American Revolution, 1916, 34.
Belva Lockwood, LOC.

Pages 52–53
Margaret N. Caine (L) and Emily S. Richards (R), in Joseph F. Smith with Church Leaders in Washington D.C., 1888, CHL.
Belva Lockwood, in *Fifty Years' Recollections*, Jeriah Bonham, 1883, 248.
Cornelia Paddock, in *The Women of Mormonism: Or, the Story of Polygamy as Told by the Victims Themselves*, Jennie Anderson Froiseth, 113, BYU.
Background: White House, The Miriam and Ira D. Wallach Division of Art, Prints and Photographs, NYPL.
Rose Elizabeth Cleveland, LOC.
May Wright Sewall, ca. 1904, National Woman's Party Records (hereafter NWP Records), LOC.
Zerelda G. Wallace, in *Woman and Temperance*, Frances E. Willard, 1883, 477.
Women's Industrial Christian Home, USHS.

Pages 54–55
Background: Salt Lake Theatre, USHS.
Memorial of the Mormon Women of Utah, Apr. 6, 1886, CHL.
Woman Suffrage in Utah, June 8, 1886, CHL.
Background: *Deseret Evening News*, Jan. 31, 1887, 3.
Grover Cleveland. LOC.
"Thanks to the Woman Suffragists," *Deseret Evening News*, Jan. 31, 1887, 3.
Sugar House Prisoners, USHS.

Pages 56–57
Background: International Council of Women: an Appeal, ca. 1888, CHL.
General Officers of the National Council of Women, LOC.
Jane S. Richards, Josephine R. West, Jane Herrick, and Josephine Herrick Preston, ca. 1896, Mary West Miller Family Photograph Collection, CHL.
International Council of Women Assembled by the National Woman Suffrage Association, 1888, CHL.
History and Minutes of the National Council of Women of the United States, ed. Louise B. Robbins, 1898, i, Schlesinger Library.
Background: Gardo House, Salt Lake City, Utah and Surrounding Area Photographs, ca. 1880–1890, CHL.
Elizabeth Lyle Saxon, in *A Woman of the Century*, Frances E. Willard, 1893, 635.
Clara Bewick Colby, Susan B. Anthony Photographs, BYU.
Mrs. Julia Ward Howe, 1899, NAWSA Scrapbooks, LOC.

Pages 58–59
The Woman's Journal, Dec. 1, 1888, 1.
Jennie Froiseth, in *The Women of Mormonism*, Jennie Anderson Froiseth, 1882, frontispiece, BYU.
Jane S. Richards, CHL.
Margaret Caine, USHS.
Woman Suffrage Association of Utah hand stamp, Church History Museum.
Emily S. Richards, courtesy of Bruce J. Nelson.

Pages 60–61
Background: Social Hall, USHS.
Elizabeth Ann Schofield [Adams], courtesy of Kristen Pieroni.
"Equal Rights Banner," BYU.
"W.S.A. in Juab County," *Woman's Exponent*, June 1, 1889.
Background and image: Sarah Staker Certificate of Membership of the Woman Suffrage Association, CHL.
Salt Lake City Assembly Hall, USHS.
Hulda Cordelia Thurston Smith, courtesy of the Morgan County Historical Society.

CHAPTER 4

Young Woman's Journal, June 1895, 390.

Pages 64–65
Background: Social Hall, Views of Early Salt Lake City, ca. 1858. CHL.
Portrait of Sarah M. Kimball, CHL.
Women Suffrage Leaders, USHS.
Elizabeth Cady Stanton and Susan B. Anthony, LOC.
Background and image: President Woodruff's Manifesto, CHL.
Maria Young Dougall, USHS.
Amalia B. Gimous Post, Wyoming State Archives Photo Collection, Wyoming State Archives.

Pages 66–67
NAWSA 23rd Annual Washington Convention program, from the collection of Ann Lewis and Mike Sponder.
"To the People of Utah," Special Collections and Archives, Merrill-Cazier Library, Utah State University.
Relief Society and National Council of Women banner, ca. 1905, Relief Society Photograph File, ca. 1920–1978, CHL.
Background and image: Utah Woman Suffrage Songbook, 1890, CHL.
Susan B. Anthony, Susan B. Anthony Photographs, BYU.
"Votes for Women" lapel pennant, LOC.

Pages 68–69

Background: *Art and Handicraft in the Woman's Building*, Maud Howe Elliott, ed., 1893.

"Looking West from Peristyle," in *Official Views Of The World's Columbian Exposition*, C. D. Arnold, H. G. Higinbotham, 1893, 17.

Utahns at Chicago World's Fair, 1893, CHL.

Martha H. Cannon photograph, 1887, CHL.

Emmeline Blanche Woodward Wells, Engraved Portrait Collection, ca. 1890, CHL.

Colorado Women Are Citizens, The Denver Public Library, Western History Collection, X-29646.

Pages 70–71

Background: *Woman's Exponent,* Aug. 1 and 15, 1894, 1.

Susan B. Anthony, LOC.

Martha Hughes Cannon, USHS.

Enabling Act, cover, 1894, Utah State Archives and Records Service (hereafter Utah State Archives).

"Equal Rights Banner," BYU.

Electa Bullock, courtesy of Dorothy Bullock Lynn.

Pages 72–73

Background: *Salt Lake Herald*, Feb. 3, 1895, 1.

Constitutional Convention (1895), USHS.

Dr. Ellen Ferguson, Engraved Portrait Collection, ca. 1890, CHL.

State Presidents and Officers of the N.A.W.S.A at Nat. Convention, 1892, Bryn Mawr.

Salt Lake City, City and County Building, USHS.

Salt Lake County Woman Suffrage Association Membership Ticket, CHL.

Utah Constitutional Convention, 1895, CHL.

Pages 74–75

Background and image: "Woman Suffragists," *Salt Lake Herald-Republican*, Mar. 19, 1895, 3.

"'Independence Day' of the Future," *Puck*, July 4, 1894, LOC.

Joanna Melton, in Grand Army of the Republic, USHS.

Brigham H. Roberts, USHS.

Woman Temple Workers, 1893, CHL.

First Presidency of the Mormon Church, Mar. 2, 1894, CHL.

Pages 76–77

Joseph F. Smith, Bathsheba W. Bigler Smith Photograph Collection, CHL.

"A Woman's Answer," *Salt Lake Tribune*, Apr. 4, 1895, 7.

Walker Opera House, USHS.

Portrait of Mary A. Burnham Freeze, CHL.

Paulina Lyman (center), in Portrait of Mary Ann Phelps Rich and Harriet Wright Phelps Holmes, CHL.

Ruth May Fox, courtesy of Brittany Chapman Nash.

Pages 78–79

Background: "Woman and the Ballot," *Salt Lake Herald-Republican*, Mar. 29, 1895, 3.

Orson F. Whitney, Orson F. Whitney Portrait Collection, CHL.

Franklin S. Richards, USHS.

Background: Constitution of the State of Utah, Preamble, May 18, 1895, Utah State Archives.

Constitution of the State of Utah, Article 4, Utah State Archives.

Pages 80–81

Susan B. Anthony with Women Suffrage Leaders, CHL.

Pages 82–83

"It Was Democracy's Day," *Salt Lake Herald-Republican*, July 14, 1895, 1.

Martha Hughes Cannon, USHS.

Sarah E. Andersen [Anderson], USHS.

"Rally of Colored Women," *Salt Lake Tribune*, Aug. 23, 1895, 3.

Background: "Utah's New Women," *The Call* (San Francisco), Sept. 14, 1895, 1.

Emma McVicker in "World of Women," *Deseret Evening News*, Oct. 13, 1900, 15.

Elizabeth Cady Stanton, LOC.

CHAPTER 5

Utah postcard, University of Northern Iowa, Palczewski Archive, University of Northern Iowa.

Pages 86–87

Background: "Woman Suffrage in Utah," *The Deseret Weekly,* Feb. 15, 1896, 1.

Statehood Celebration 1896 [ZCMI], USHS.

Salt Lake Tabernacle—Organ, USHS.

Background: "Populist State Convention," *Salt Lake Herald-Republican*, June 21, 1896, 7.

Susa Young Gates, "Where Women Vote. In Utah," NAWSA Political Equality Series, 1899, CHL.

Amanda Knight, USHS.

Women in front of house, Alma W. Compton glass plate negative collection, item 122, CHL.

Pages 88–89

Background: "Sample Ballot," *Salt Lake Herald-Republican*, Oct. 31, 1896, 10.

Martha Hughes Cannon with Daughter, Mattie, USHS.

Legislature—Senate, 1897, USHS.

Background: Suffrage map with 4 states, in *History of Woman Suffrage vol. 6, 1900–1920,* ed. Ida Husted Harper, 1922, 626.

Eurithe K. LaBarthe, USHS.

Woman Suffrage Button, NMAH.

Woman Suffrage Flag, ca. 1900, NMAH.

Pages 90–91

Background: Martha Hughes Cannon, "Woman Suffrage in Utah," Feb. 15, 1898, CHL.

Martha Hughes Cannon, used by permission, USHS.

The Woman's Exponent, Jan. 15 and Feb. 1, 1897, 1.

The Apotheosis of Suffrage, 1896, LOC.

Background: Art Bill, 1899 Session, Utah Senate Bill 86, Utah State Archives.

Alice Merrill Horne, USHS.

Elizabeth Taylor in *Utah Plain Dealer*.

"Echoes of the Election," *Broad Ax*, Nov. 12, 1898, 1.

Pages 92–93

Background: May Wright Sewall to Susa Young Gates, February 1, 1899, Susa Young Gates Papers, CHL.

Hannah Kaaepa, CHL.

"Utah Women at Washington," *Salt Lake Herald-Republican*, Feb. 26, 1899, 4.

International Congress of Women, 1914, CHL.

Background: Public Health Bill, 1899 Session, Utah Senate Bill 40, Utah State Archives.

Voting Record for Utah Senate Bill 40, 1899 Session, Utah State Archives.

Art Bill, 1899 Session, Utah House Bill 124, Utah State Archives.

Daffodil Place Card, 1910, NAWSA Scrapbooks, LOC.

Pages 94–95

Mrs. Carrie Chapman Catt, NWP Records, LOC.

"Editorial Notes," *The Woman's Exponent*, Nov. 1, 1899, 69.

"Delegates to Go to London," *Salt Lake Herald-Republican*, June 7, 1899, 7.

Carrie Chapman Catt, in *American Women: Fifteen Hundred Biographies*, Frances E. Willard and Mary A. Livermore, 1897, 162.

Background: Susan B. Anthony Eightieth Birthday Celebration, 1900, NAWSA Scrapbooks, LOC.

Susan B. Anthony's black silk dress, image used with the permission of the National Susan B. Anthony Museum and House, Rochester, NY.

Silk culture, ca. 1895, CHL.

Susan B. Anthony Eightieth Birthday Celebration, 1900, NAWSA Scrapbooks, LOC.

Pages 96–97

Emma McVicker, Salt Lake Tribune files.

Herald-Republican, Exterior with Girls on Porch, USHS.

First Presidency, ca. 1900, CHL.

Background: First International Woman Suffrage Conference, 1902, NAWSA Scrapbooks, LOC.

Official representatives of the International Council of Women, 1902, LOC.

Three unidentified women at a booth at the MWSA bazaar, ca. 1914, Schlesinger Library.

Group portrait of pages at an IWSA conference in Stockholm, 1911, Schlesinger Library.

National American Woman Suffrage Association Delegate ribbon, LOC.

Pages 98–99

Background: *International Congress of Women, Berlin, 1904*, booklet of participants, Charlotte Perkins Gilman Papers, Schlesinger Library.

Mary G. Coulter in "Only Woman in the Legislature Made Chairman of Important Committee," *Salt Lake Herald-Republican*, Jan. 14, 1903, 8.

International Council of Women luncheon, 1904, Archiv der deutschen Frauenbewegung, Kassel.
Lydia Alder and Corrine Allen, in *International Congress of Women, Berlin, 1904*.
Background: Saltair, Views of Salt Lake City, Utah, ca. 1870s–early 1900s, CHL.
"Lifting as We Climb" banner, Collection of the Smithsonian National Museum of African American History and Culture.
Elizabeth Taylor, *Utah Plain Dealer*, Aug. 4, 1900.
The Phyllis Wheatley Club, Buffalo, New York, LOC.
Headquarters for Colored Women Voters, ca. 1920, Schomburg Center for Research in Black Culture, NYPL.

Pages 100–101
Background: *Deseret Evening News*, Mar. 17, 1906, 1.
Susan B. Anthony, LOC.
"I Wonder If It's Really Becoming?" *The Abbeville Press and Banner,* Mar. 25, 1908, 2.
"Utah Woman Pleads for Women Suffragists," *Salt Lake Tribune,* Mar. 4, 1908, 1.
Background: "Mrs. Lucy Clark Talks to Women of Chicago," *Inter-Mountain Republican,* June 18, 1908, 7.
Lucy A. Clark, courtesy of Lori Billings Smith.
Osborne Auto Party, USHS.
Elizabeth Pugsley Hayward, courtesy of the Wessman family.

Pages 102–3
Background: Instructions to Workers for the Woman Suffrage Petition to Congress, ca. 1909, NAWSA Scrapbooks, LOC.
"International Doll Exhibit for Congress Funds," 1913, NYPL Digital Collections.
"Votes for Women" ribbon, 1911, NAWSA Scrapbooks, LOC.
"Votes for Women" stamp, 1910, NAWSA Scrapbooks, LOC.
An Early Political Parade in Michigan, Bryn Mawr.
"Uncle Sam's Newest Girl Baby," University of Washington Libraries, Special Collections, UW28335z.
Relief Society Presidency and General Board, 1914, CHL.
Posting signs to promote woman suffrage, Washington Equal Suffrage Association, Seattle, 1910, University of Washington Libraries, Special Collections, A. Curtis 19943.
"Hurrah for a Free West!!" *Western Woman Voter,* Sept. 1911, 12.

CHAPTER 6
Suffragists lobby Senator Smoot, NWP.

Pages 106–7
Background: "Women vote for President . . . why not in California?" UC Berkeley, Bancroft Library.
Susa Young Gates, USHS.
"The Steamroller," LOC.
Kanab Ladies' Town Board, USHS.
"Meanwhile the Ladies Have Been Having a Perfectly Lovely Time," *The Woman's Journal,* Nov. 30, 1912, 1, courtesy Emmett D. Chisum Special Collections.
Illinois suffrage cartoon, *The Commoner,* Aug. 1, 1913, 20.
Florence Allen holding flag at Ohio Woman Suffrage Headquarters, 1912, LOC.
Margaret Zane Cherdron, NWP Records, LOC.

Pages 108–9
Background: "Mrs. Groshell Leads Big Parade," *Salt Lake Herald-Republican* Mar. 10, 1913, 3.
Head of Suffrage Parade, Mar. 13, 1913. LOC.
Suffrage Hikers Collecting, 1913, LOC.
Official program, Woman Suffrage Procession, Mar. 3, 1913, LOC.
Suffragists in Parade, LOC.
Background: *Woman's Exponent*, Feb. 1, 1914, 100.
Procession of 2,000 Women Voters, 1915, NWP Records, LOC.
Margaret Vale in Suffrage Parade, LOC.
George Sutherland, 1916, LOC.

Pages 110–11
Alice Paul, NWP Records, LOC.
Congressional Union Summer Headquarters, LOC.
Parade of Utah women voters on Main Street, NWP.
Emily Perry speaking in Salt Lake City, 1916, NWP.
Suffrage envoys greeted in New Jersey, 1915, LOC.

Pages 112–13
Background: Map of Envoys' Route, 1916, NWP Records, LOC.
Senator Sutherland and suffragists greeting envoy, 1915, NWP.
Sara Bard Field speaking at Salt Lake City, 1915, NWP Records, LOC.
Welcome to Envoys of Women Voters, 1915, LOC.
Suffragettes [Suffragists] at Capitol, LOC.
Jeanette Young Easton, USHS.
Alice Reynolds (R) in "Suffrage Special," 1916, NWP.
Touring the State for Suffrage, Bryn Mawr.

Pages 114–15
Background: "Votes for Women" poster, UC Berkeley, Bancroft Library.
Women's Voter Convention, Sept. 1915, NWP Records, LOC.
NAWSA Suffrage Headquarters, in *History of Woman Suffrage vol. 5, 1900–1920*, ed. Ida Husted Harper, 1922, 632.
National Woman's Party Utah Headquarters, NWP.
Background: Democratic Women, 1914, LOC.
Inez Milholland, LOC.
Zitkála-Šá, 1898, NMAH.
Grace Stratton Airey, Woman Legislator of Utah, USHS.
Zitkála-Šá, 1898, National Museum of American History, courtesy of Mina Turner.

Pages 116–17
Pennsylvania Suffragists on the Picket Line, LOC.
First Suffrage Picket Line, January 10, 1917, NWP Records, LOC.
Picketing in All Sorts of Weather, 1917, NWP Records, LOC.
"The Weaker Sex?" *Puck*, Nov. 7, 1914, 5, LOC.
Background: War Relief Parade in Ogden, Edna C. Hammon Crossley Photograph Collection, CHL.
Maud Fitch, USHS.
"First Ambulance on Duty," in *Women and War Work*, Helen Fraser, 1918, 56.
New York Woman Suffrage Broadside, Ontario County (NY) Historical Society.
Lily Wolstenholme, courtesy of Daniel A. Wolstenholme, Jr.

Pages 118–19
Minnie [Mrs. R. B.] Quay, NWP Records, LOC.
Lovern Robertson (4th from L) in Picket Line of Nov. 10, 1917, NWP Records, LOC.
Miss Lucy Burns in Occoquan Workhouse, NWP Records, LOC.
Lovern Robertson, NWP Photograph Collection, LOC.
National American Suffrage Congress, 1919, LOC.
Suffrage Parade with "President Wilson says" sign, authors' collection.
NAACP Silent Protest Parade, 1917, LOC.

Pages 120–21
Background: "Special Legislature for Suffrage Asked," *Salt Lake Tribune*, July 7, 1919, 22.
"Votes for Women Bandwagon," 1918, LOC.
Dr. Anna Howard Shaw, 1919, LOC.
Conference of Governors, Utah, 1920, NWP.
Background: Hotel Utah, Views of Salt Lake City and Utah, c. 1915–1923, CHL.
State Senator Elizabeth A. Hayward, Bryn Mawr.
Woodrow Wilson visits Salt Lake City, 1919, CHL.
Twelfth Session of Utah State House of Representatives, 1917, CHL.
Carrie Chapman Catt on telephone, LOC.

Pages 122–23
Background: National American Women Suffrage Convention, St. Louis, 1919, LOC.
Alice Paul unfurling ratification banner, 1920, NWP Records, LOC.
Women's Conventions in Chicago, 1920, CHL.
Febb Burn, Harry T. Burn Papers, C.M. McClung Historical Collection, Knox County Public Library.
"At Last," *The Suffragist*, June 21, 1919, Bryn Mawr.
Suffragists celebrate, Bryn Mawr.
Three suffragists casting votes in New York City, LOC.
Women out in force, LOC.

EPILOGUE
Ruth May Fox, 1956, USHS.

IN MEMORIAM
Deborah Ann Coulam Wheelwright, courtesy of the Wheelwright family.